Statistical Methods

Statistical Methods

Mujahida Sayyed
Assistant Professor
College of Agriculture
Jawahrlal Nehru Krishi Vishwavidayala
Ganj Basoda, Madhya Pradesh

NEW INDIA PUBLISHING AGENCY
Pitam Pura, New Delhi – 110 088

New India Publishing Agency
101, Vikas Surya Plaza, CU Block, LSC Market
Pitam Pura, New Delhi 110 034, India
Phone: + 91 (11)27 34 17 17 Fax: + 91(11) 27 34 16 16
Email: info@nipabooks.com
Web: www.nipabooks.com

Feedback at feedbacks@nipabooks.com

ISBN : 978-93-58870-09-1

Composed and Designed by NIPA

Preface

Statistics is used in two senses, singular and plural. In the singular, it concerns with the whole subject of statistics, as a branch of knowledge. In the plural sense, it relates to the numerical facts, data gathered systematically with some definite object in view. Thus, Statistics is the science, which deals with the collection, analysis and interpretation of data.

An understanding of the logic and theory of statistics is essential for the students of agriculture who are expected to know the technique of analyzing numerical data and drawing useful conclusions. It is the intention of the author to keep the practical manual at a readability level at appropriate for students who do not have a mathematical background.

This manual has been prepared for the students and teachers as well to acquaint the basic concepts of statistical principles and procedures of calculations as per the syllabi of 5th Dean's committee of ICAR for undergraduate courses in agriculture and allied sciences. I wish the practical manual would be very much useful for students and teachers.

The author wishes to thank the New India Publishing Agency, New Delhi for its cooperation in all aspects. Further suggestions from teachers and students for the improvement of this practical manual would be gratefully acknowledged.

Mujahida Sayyed

Contents

1

Construction of Frequency Distribution

Frequency Distribution: A tabular presentation of the data in which the frequencies of values of a variable are given along with them is called a frequency distribution. Two types of frequency distribution are available

1. Discrete Frequency Distribution or Ungrouped Frequency Distribution
2. Continuous Frequency Distribution or Grouped Frequency Distribution

Objective (a): Prepare a discrete frequency distribution from the following sentence in English: "Today there is hardly a phase of Endeavour which does not find statistical devices at least occasionally useful."

Theory: This problem relates to discrete frequency distribution i. e. a frequency distribution which is formed by distinct values of a discrete or continuous variable.

Process: For Discrete Frequency Distribution

Step I. Set the data in ascending order.

Step II. Make a blank table consisting of three columns with the title: Variable, Tally Marks and Frequency.

Step III. Read off the observations one by one from the data given and for each one record a tally mark against each observation.

Step IV. In the end, count all the tally marks in a row and write their number in the frequency column.

Step V. Write down the total frequency in the last row at the bottom.

Flow Diagram:

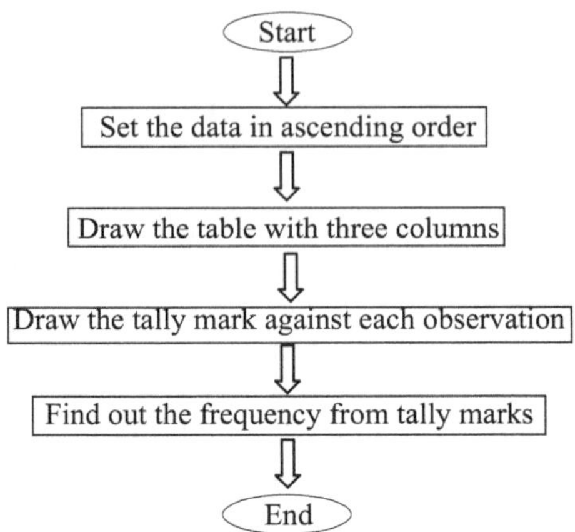

Calculation: First of all write the number of letters in each word as given below

5	5	2	6	1	5	2	9	5	4
3	4	11	7	2	5	12	6		

Arrange data in ascending order

1	2	2	2	3	4	4	5	5	5
5	5	6	6	7	9	11	12		

Counting the numbers by tally method we get the required discrete frequency distribution:

No. of Letters Variable (X)	**Tally Marks**	**No. of Words Frequency(f)**
1	|	1
2	|||	3
3	|	1
4	||	2
5	~~||||~~	5
6	||	2
7	|	1
9	|	1
11	|	1
12	|	1
Total		**18**

Result: The required discrete frequency distribution is

No. of Letters Variable (X)	Tally Marks	No. of Words Frequency(f)
1	|	1
2	|||	3
3	|	1
4	||	2
5	~~||||~~	5
6	||	2
7	|	1
9	|	1
11	|	1
12	|	1
Total		**18**

Objective (b) : 20 students appear in an examination. The marks obtained out of 50 maximum marks are as follows:

5, 16, 17, 17, 20, 21, 22, 22, 22, 25, 25, 26, 26, 30, 31, 31, 34, 35, 42 and 48.

Prepare a frequency distribution taking 10 as the width of the class-intervals .

Theory: This problem relates to a continuous frequency distribution i.e. a frequency distribution which obtained by dividing the entire range of the given observations on a continuous variable into groups and distributing the frequencies over these groups . It can be done by two methods

1. Inclusive method of class intervals: When lower and upper limit of a class interval are included in the class intervals.
2. Exclusive method of class intervals: When the upper limit of a class interval is equal to the lower limit of the next higher class intervals.

Process: For Continuous Frequency Distribution

Step I. Set the data in ascending order.

Step II. Classify the data by exclusive and/or inclusive method for the desired width of the class intervals.

Step III. Make a blank table consisting of three columns with the title: Variable, Tally Marks and Frequency.

Step IV. Read off the observations one by one from the data given and for each one record a tally mark against each observation.

Step V. In the end, count all the tally marks in a row and write their number in the frequency column.

Step VI. Write down the total frequency in the last row at the bottom.

Flow Diagram:

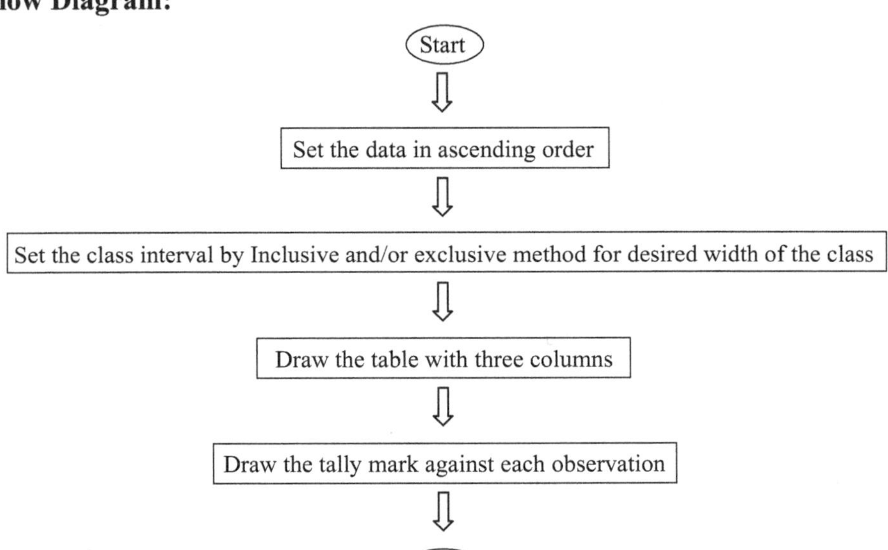

Calculation: Arrange the data in the ascending order

5	16	17	17	20	21	22	22	22	25
25	26	26	30	31	31	34	35	42	48

Here lower limit is 5 and upper limit is 48.

Since it is given that the desired class interval is 10, so frequency distribution for Inclusive Method of Class intervals:

Marks	Tally Marks	No. of students
1-10	|	1
11-20	||||	4
21-30	~~||||~~ ||||	9
31-40	||||	4
41-50	||	2
Total		**20**

Exclusive Method of Class intervals:

Marks	Tally Marks	No. of students
0-10	|	1
10-20	|||	3
20-30	卌 ||||	9
30-40	卌	5
40-50	||	2
Total		**20**

Result: we get the required frequency distribution for Inclusive and Exclusive method of class intervals

Marks	Tally Marks	No. of students
1-10	|	1
11-20	||||	4
21-30	卌 ||||	9
31-40	||||	4
41-50	||	2
Total		**20**

Marks	Tally Marks	No. of students
0-10	|	1
10-20	|||	3
20-30	卌 ||||	9
30-40	卌	5
40-50	||	2
Total		**20**

Related Questions:

Q.1 What is distribution?

Q.2 What is frequency?

Q.3 What is discrete variable?

Q.4 What is continuous variable?

Q.5 What is exclusive and inclusive method of class intervals?

2

Construction of Bar Diagram and Ogive Curve

The important convincing appealing and easily understood method of presenting the statistical data is the use of diagrams and graphs. Diagram makes the comparison of the data but graphs makes the relationship among the data.

Objective: Aggregated figures for merchandise export in India for eight years are as follows

Years	1971	1972	1973	1974	1975	1976	1977	1978
Exports (million Rs.)	1962	2174	2419	3024	3852	4688	5555	5112

For the above data draw Bar diagram and Ogive curve

Theory:

Simple Bar Diagram: The simple bars are the thick or thin lines without any subdivision with equal width whose lengths are proportional to the given figures. They are drawn at equal distances. The width and the distance are not taken into consideration but they are simply to make the diagram attractive.

Process: For Simple Bar Diagram

Step I: Draw X and Y axis.

Step II: Take year on X axis .

Step III: Take scale of 1000 on Y axis which represent Exports.

Step IV: Draw the equal width bars on X axis.

Flow Diagram:

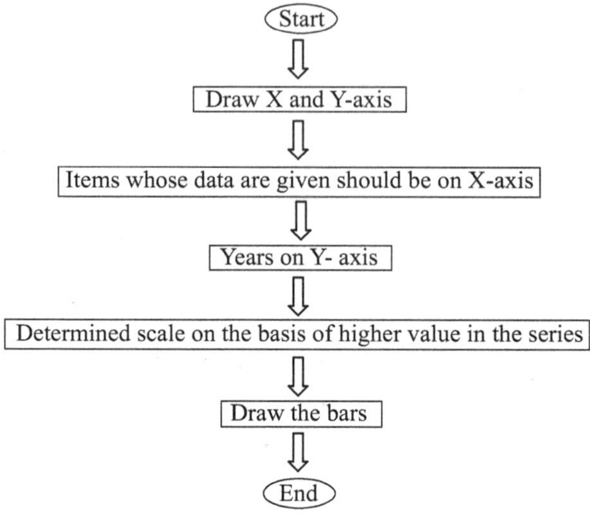

Process: Simple Bar Diagram:

Year	Export
1971	1962
1972	2174
1973	2419
1974	3024
1975	3852
1976	4688
1977	5355
1978	5112

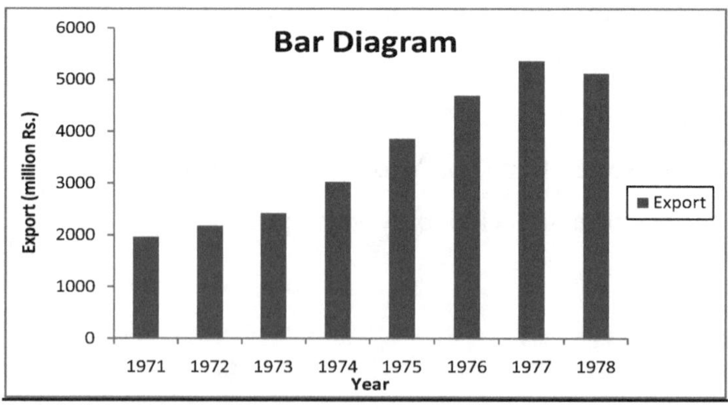

Ogive Curve: The cumulative distribution may be represented by a smooth curve through the extremities of the ordinates representing the successive cumulative frequencies. Such a smooth curve with an elongated S shape is called cumulative frequency curve. Also from its resemblance to the shape of a moulding in architecture known as an ogee, this curve is called an Ogive.

Process For Ogive Curve

Step I: Prepare the Cumulative frequency less than (by adding frequencies) and more than types (by subtracting frequencies)

Step II: Mark the class intervals along the X axis.

Step III: Plot the corresponding cumulative frequency of less than type along against the value of upper limits of the class intervals on the Y axis.

Step IV: Plot the corresponding cumulative frequency of more than type against the value of the lower limit of the class interval on the Y axis.

Step V: Join the plotted points by means of free hand curve separately for less than and more than cumulative frequencies.

Step VI: The curve so drawn is called Ogive curve.

Flow Diagram

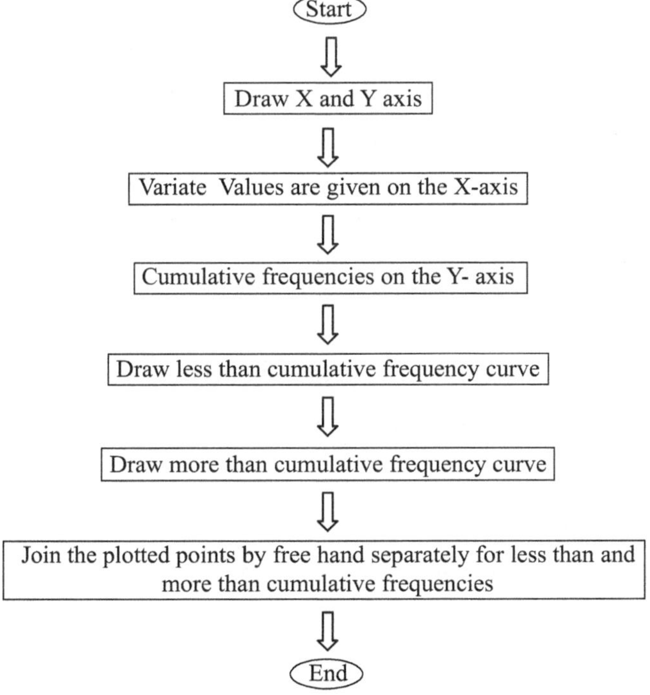

Calculation:

Year	Export	cf less than	cf more than
1971	1962	1962	26624
1972	2174	4136	24450
1973	2419	6555	22031
1974	3024	9579	19007
1975	3852	13431	15155
1976	4688	18119	10467
1977	5355	23474	5112
1978	5112	28586	0

Ogive Curve:

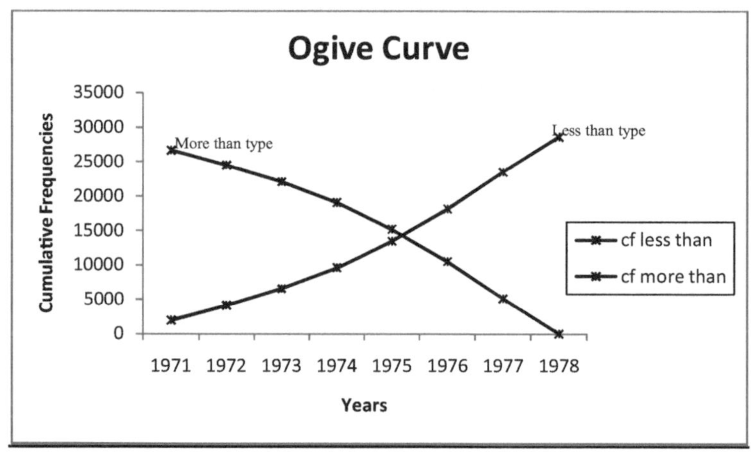

Result: The above figures shows the Bar diagram and Ogive curve.

Related Questions:

Q.1 What is the difference between graph and diagram?

Q.2 In a Bar diagram X and Y axis represents?

Q.3 What is an ogive curve?

Q.4 For which purpose ogive curve is used?

Q.5 What do you mean by cumulative frequency?

3

Construction of Pie Chart

Pie charts are useful to compare different parts of a whole amount. They are often used to present financial information. E.g. A farmer's expenditure on farm can be shown to be the sum of its parts including different expense categories such as cultivation, irrigation, sowing, harvesting, and general running costs (i.e. rent, electricity, fuel etc).

A pie chart is a circular chart in which the circle is divided into sectors. Each sector visually represents an item in a data set to match the amount of the item as a percentage or fraction of the total data set.

Objective: A family's weekly expenditure on its house mortgage, food and fuel is as follows:

Expense	**Rupees**
Mortage	300
Food	225
Fuel	Fuel

Draw a pie chart to display the information.

Process:

Step I:

The total weekly expenditure = 300+225+75
= 600Rs.

Step II:

Percentage of weekly expenditure on:

Mortage $= \frac{300}{600} * 100\% = 50\%$

Food $= \frac{225}{600} * 100\% = 37.5\%$

Fuel $= \frac{75}{600} * 100\% - 12.5\%$

Step III: To draw a pie chart, divide the circle into 100 percentage parts. Then allocate the number of percentage parts required for each item.

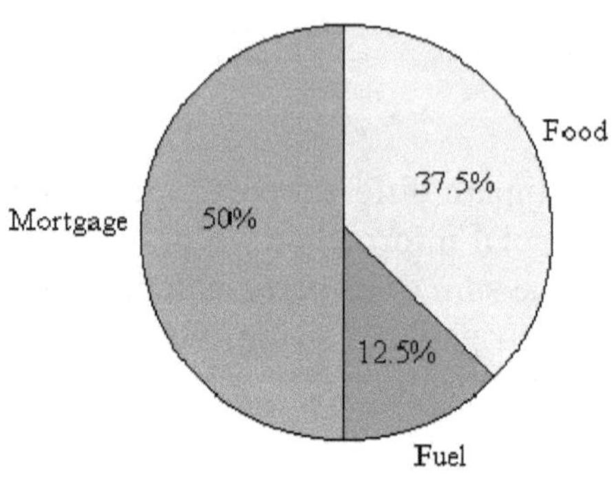

Flow Diagram

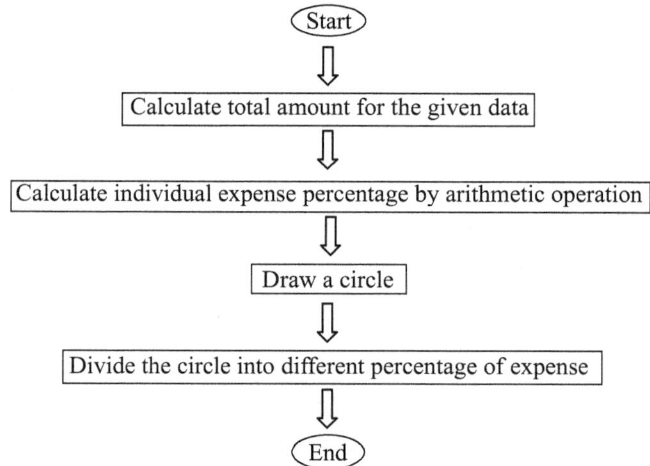

Result: Above figure shows the pie diagram of the given data.

Related Questions

Q.1 What is diagram?

Q.2 What is the importance of pie diagram?

Q.3 What is the angle of a circle?

Q.4 What is acute angle?

Q.5 What is obtuse angle?

4

Study of Mean, Median and Mode for Ungrouped Data

Measures of Central Tendency

The term 'Mean" or 'average' is something, we have been familiar with from a very early age when we start analyzing our marks on report card. We add together all of our test results and then divide it by the sum of the total number of marks there are. We often call it the average. However, statistically it's the Mean!

The Median is the 'middle value'. When the totals of the list are odd, the median is the middle entry in the list after sorting the list into increasing order. When the totals of the list are even, the median is equal to the sum of the two middle (after sorting the list into increasing order) numbers divided by two. Thus, remember to line up values, the middle number is the median!

The Mode in a list of numbers refers to the list of numbers that occur most frequently. A trick to remember this one is to remember that mode starts with the same first two letters that most does. Most frequently - Mode.

For Ungrouped Data

Objective (a): Find the mean of four tests results: 15, 18, 22, 20.

Theory: If $X_{1,}$ $X_{2,}$ $X_{3,\dots}$ X_n are the n observations, their mean is given by and define as follows

$$\bar{X} = \frac{\text{Sum of observations}}{\text{Total no.of observations}}$$

$$= \frac{X_1 + X_2 + X_3 + \dots + X_n}{n}$$

$$= \frac{\sum_{i=1}^{n} X_i}{n}$$

Process: Step I: The sum of observation is: 75

Step II : Total no. of observation is: 4

Step III : Then mean is calculated by

$$\bar{X} = \frac{75}{4}$$

$$\bar{X} = 18.75$$

The 'Mean' (Average) is 18.75 .

Flow Diagram

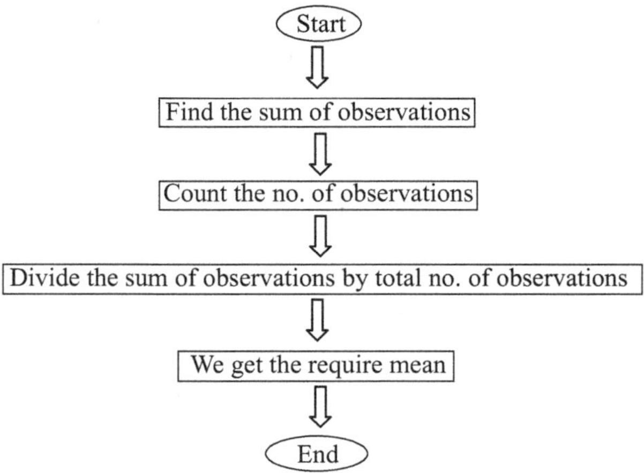

Result : The mean of the given test result is 18.75.

Objective (b): Find the Median of 9, 3, 44, 17, 15 and 8, 3, 44, 17, 12, 6.

Theory: After arranging the data in ascending or descending order the middle term of the data is known as median, if the data are odd in number. But, if the data is even in number than the mean of the middles terms become median.

Process: (Odd amount of numbers)

Step I: Arrange the data in ascending order: 3, 9, 15, 17, 44.

Step II: The middle term is 15, so the Median is: 15 (Even amount of numbers)

Step I : Arrange the data in ascending order : 3, 6, 8, 12, 17, 44

Step I : The middle numbers are 8 and 12. The mean of 8 and 12 is

8 + 12 = 20 ÷ 2 = 10, so the Median is 10.

Flow Diagram

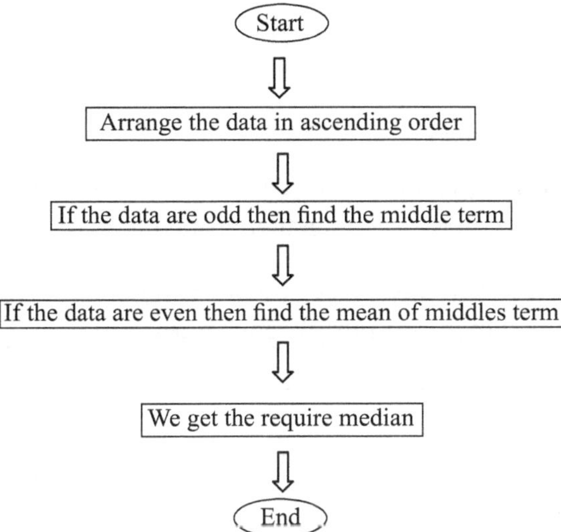

Result: The median for odd amount of the given data is 15 and for even amount of the data is 10.

Objective (c): Find the mode of: 9, 3, 3, 44, 17 , 17, 44, 15, 15, 15, 27, 40, 8.

Theory: The number has the higher frequency becomes mode of the given data.

Process: Step I: Put the numbers in increasing order:

3, 3, 8, 9, 15, 15, 15, 17, 17, 27, 40, 44, 44.

Step II: 15 occur the most at 3 times,

Step III: Mode is 15.

Flow Diagram

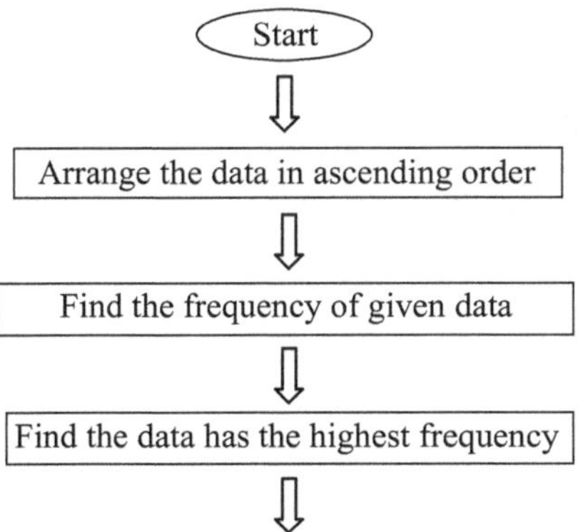

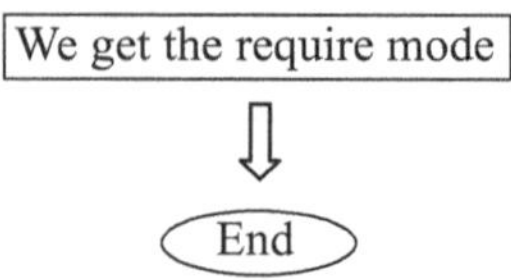

Result: For the given data mode is 15.

*It is important to note that there can be more than one mode and if no number occurs more than once in the set, then there is no mode for that set of numbers.

What will happen to the measures of central tendency if we add the same amount to all data values, or multiply each data value by the same amount?

	Data	Mean	Mode	Median
Original Data Set:	6, 7, 8, 10, 12, 14, 14, 15, 16, 20	12.2	14	13
Add 3 to each data value	9, 10, 11, 13, 15, 17, 17, 18, 19, 23	15.2	17	16
Multiply 2 times each data value	12, 14, 16, 20, 24, 28, 28, 30, 32, 40	24.4	28	26

When added: Since all values are **shifted** the same amount, the measures of central tendency all shifted by the same amount.

When multiplied: Since all values are affected by the same multiplicative values, the measures of central tendency will feel the same affect. If you multiply each data value by 2, you will multiply the mean, mode and median by 2.

Related Questions

Q.1 What is mean?

Q.2 What is median?

Q.3 What is mode?

Q.4 What is the relation among mean, median and mode?

Q.5 Which is the best measure of central tendency and why?

5

Study of Mean, Median and Mode for Grouped Data

Objective: The distribution of age of males at the time of marriage was as follows

Age (Years)	18-20	20-22	22-24	24-26	26-28	28-30
No. of Males	5	18	28	37	24	22

Find at the time of marriage

1. The Average age
2. The Modal age
3. The Median age

Theory:

i. The Average age i.e.

Mean = $A + \frac{\sum fd}{N}$

where

A = Assumed Mean

d = (X-A)/h

N = sum of frequencies

ii. The Modal age

Mode = $\frac{(f_1 - f_0)}{(2f_1 - f_0 - f_2)} h$

Where

f_1 = frequency of mode class

f_0 = frequency of class preceding the mode class

f_2 = frequency of class succeeding the mode class

h = class interval

iii. The Median age

Median = $l+\left(\frac{N}{2}-c\right)\frac{h}{f}$

Where

l = lower limit of median class

c = cumulative frequency just greater than N/2.

h = class interval

f = frequency of median class

Process

(i) For Average Age:

Step I: Let A= 23

Step II: Table

Age	No. of Males f	x	d = (x-A)/h	fd
18-20	5	19	-2	-10
20-22	18	21	-1	-18
22-24	28	23	0	0
24-26	37	25	1	37
26-28	24	27	2	48
28-30	22	29	3	66
Total	134			123

Step III: According to formula

Mean = $A+\frac{\sum fd}{N}$

= 23 + 123/134 = 24.8358

Step IV: Average Age = 24.83582

Flow Diagram

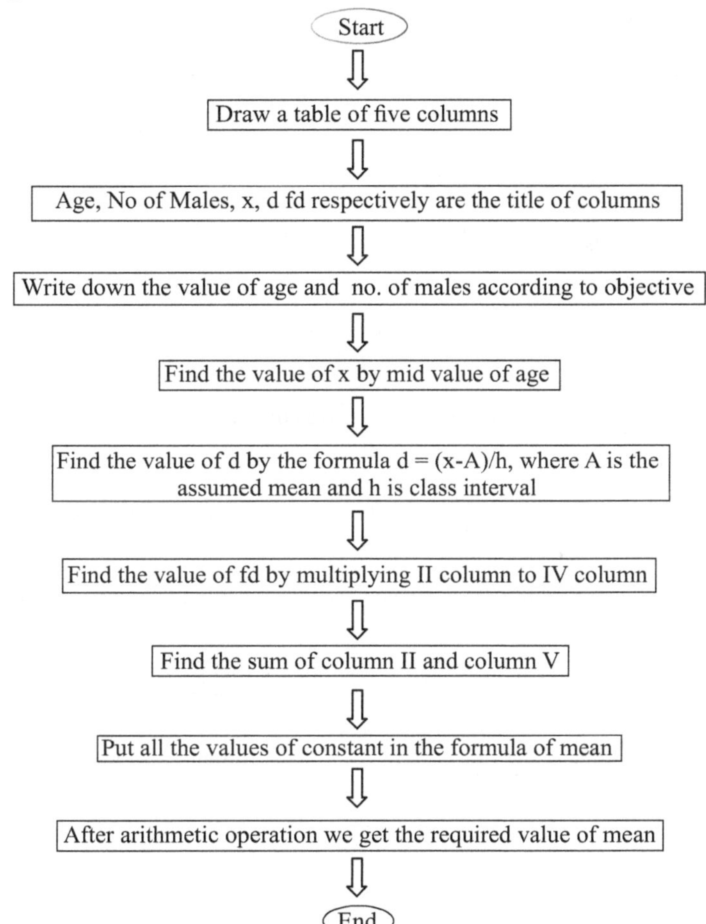

(ii) For Modal Age

Step I: Table

Age	No. of males f
18-20	5
20-22	18
22-24	28
24-26	37
26-28	24
28-30	22
Total	**134**

Step II: Here highest frequency is 37 hence the mode class is 24-26 then according to formula

$f_1 = 37$, $f_0 = 28$ and $f_2 = 24$

$$\text{mode} = \frac{(37-28)}{(2*37-28-24)}*2 = 24.4091$$

Step III: Modal Age = 24.4091

Flow Diagram:

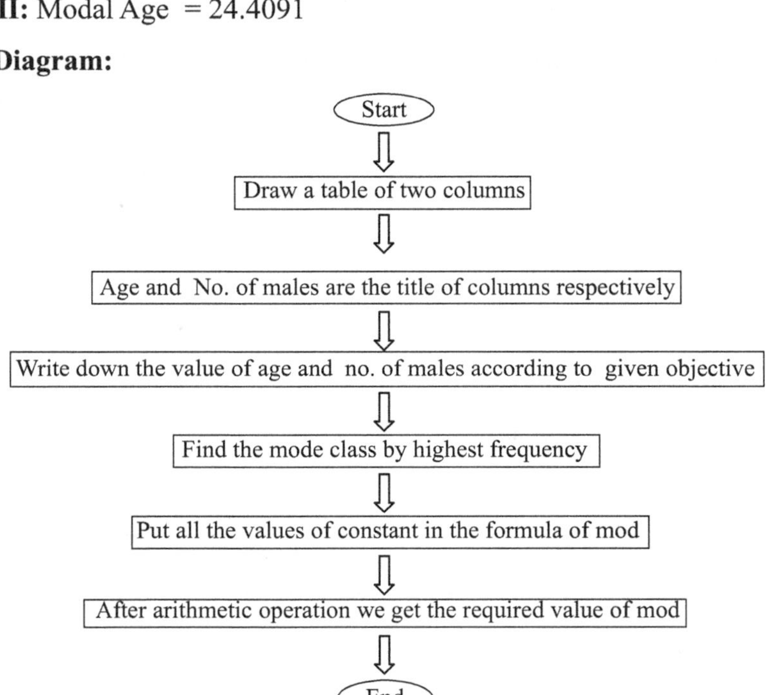

(iii) For Median age

Step I: Table

Age	No. of males f	X	cf
18-20	5	19	5
20-22	18	21	23
22-24	28	23	51
24-26	37	25	88
26-28	24	27	112
28-30	22	29	134
Total	**134**		

Step II: here N/2 = 67

The cumulative frequency just greater than 67 is 88.

Hence the median class is 24-26 .

Step III: According to formula here

l = 24, N/2 = 67, c=51, h=2 and f=37

$$\text{Median} = 24 + (67 - 51) * \frac{2}{37} = 24.8649$$

Step IV: Median Age = 24.8649

Flow Diagram:

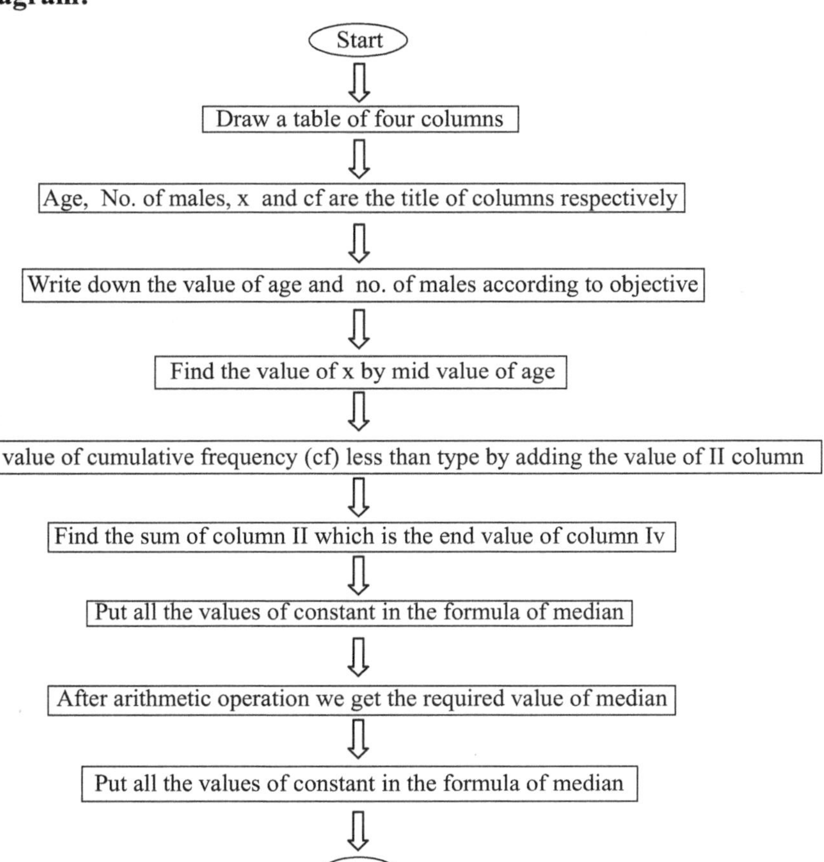

Result: From The given data we get

(i) Average age = 24.8358

(ii) Modal Age = 24.4091

(iii) Median Age = 24.8649

Related Questions:

Q.1 What is the use of mean?

Q.2 Where median is suitable for use?

Q.3 Where mode is suitable for use?

Q.4 What is cf?

Q.5 What do you mean by grouped and ungrouped series?

6

Study of Quartile, Deciles and Percentile Ungrouped and Grouped Data

Paratition Value

Median is calculated for dividing the series into two equal parts. If the series is divided into four, eight, ten or hundred parts then the measures are called Quartile, Octile, Decile and Percentile respectively denoted by $Q_{(1,2,3)}$, $O_{(1,2\cdots7)}$, $D_{(1,2\ldots9)}$ and $P_{(1,2\ldots99)}$.

Quartile:

Q_n = value of the $\left(n*(\frac{N+1}{4})^{th}\right)$ *item*

Where n= 1,2 and 3

Octile

O_n = value of the $\left(n*(\frac{N+1}{8})^{th}\right)$ *item*

Where n = 1,2 ...7.

Decile

D_n = value of the $\left(n*(\frac{N+1}{10})^{th}\right)$ *item*

Where n= 1,2 ...9.

Percentile

P_n = value of the $\left(n*(\frac{N+1}{100})^{th}\right)$ *item*

Where n= 1,2 … 99.

FOR UNGROUPED DATA

Objective (a): Following data shows the marks of Statistics, find the value of 3rd quartile, 5th Decile and 25th percentile 32 33 47 55 68 75 80.

Theory

Third Quartile:

Q_3 = value of the $\left(3*(\frac{N+1}{4})^{th}\right) item$

Fifth Decile:

D_5 = value of the $\left(5*(\frac{N+1}{10})^{th}\right) item$

Percentile:

P_{25} = value of the $\left(25*(\frac{N+1}{100})^{th}\right) item$

Process: **Step I:**

S. No.	X
1	32
2	33
3	47
4	55
5	68
6	75
7	80

Step II:

Third Quartile:

Q_3 = value of the $\left(3*(\frac{N+1}{4})^{th}\right) item$

= value of the $\left(3*(\frac{7+1}{4})^{th}\right) item$

= value of the (3*2) *item*

= value of the (6th) *item*

$Q_3 = 75$

Step III

Fifth Decile:

D_5 = value of the $\left(5*(\frac{N+1}{10})^{th}\right)$ *item*

= value of the $\left(5*(\frac{7+1}{10})^{th}\right)$ *item*

= value of the $(5*(0.8)^{th})$ *item*

= value of the (4th) *item*

$D_5 = 55$

Percentile:

P_{25} = value of the (25* $\left(\frac{N+1}{100})^{th}\right)$ *item*

= value of the (25* $\left((\frac{7+1}{100})^{th}\right)$ *item*

= value of the $(25*(0.08)^{th})$ *item*

= value of the (2nd) *item*

= 33

Flow Diagram

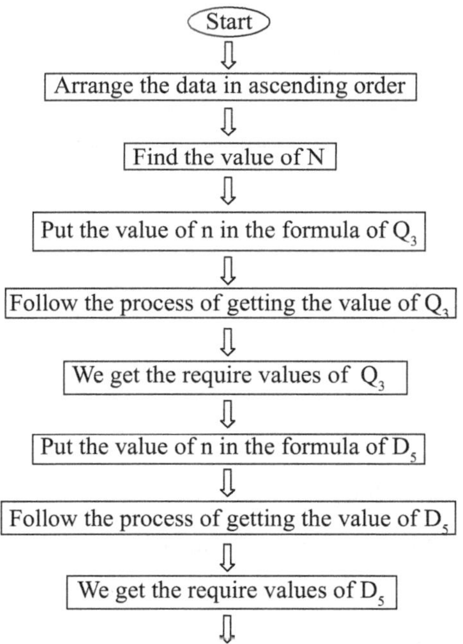

Continued

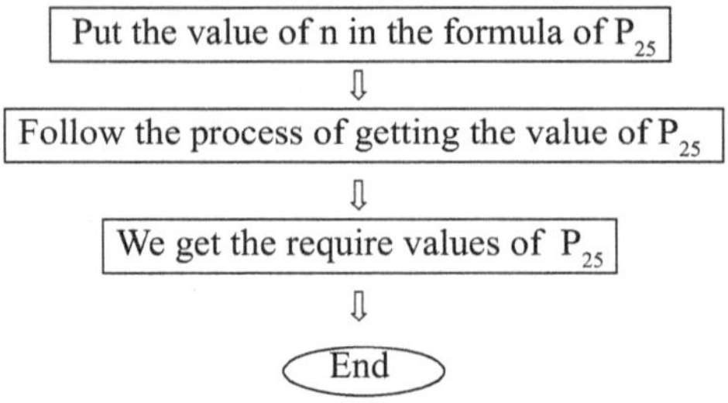

Put the value of n in the formula of P_{25}

⇩

Follow the process of getting the value of P_{25}

⇩

We get the require values of P_{25}

⇩

End

Result: From the data given we get the following value

(i) $Q_3 = 75$

(ii) $D_5 = 55$

(iii) $P_{25} = 33$

FOR GROUPED DATA

Objective (b): From the following data calculate quartile , 8th Decile and 90th percentile.

Marks	No. of Student
Below 5	14
Below 10	40
5-15	76
15 and above	110
20-25	40
25 and above	10
30 and above	2

Theory:

First Quartile:

q_1 = value of the $(\frac{N+1}{4})^{th}$ *item*

$$Q_1 = l + (q_1 - c)\frac{h}{f}$$

Where

l = lower limit of median class

c = cumulative frequency just greater than N/2.

h = class interval

f = frequency of median class

Third Quartile:

q_3 = value of the $\left(3*(\frac{N+1}{4})^{th}\right) item$

$Q_3 = l+(q_3-c)\frac{h}{f}$

Eighth Decile:

d_8 = value of the $\left(8*(\frac{N+1}{10})^{th}\right) item$

$D_8 = l+(d_8-c)\frac{h}{f}$

Ninetieth Percentile:

p_{90} = value of the $\left(90*(\frac{N+1}{100})^{th}\right) item$

$P_{90} = l+(p_{90}-c)\frac{h}{f}$

Process: Step I: Table

Marks	No. of Students	cf
0-5	14	14
5-10	(40-14)=26	40
10-15	(76-26) = 50	90
15-20	(110-(40+10)) = 60	150
20-25	40	190
25-30	(10-2) = 8	198
30-3.5	2	200

Step II: First Quartile:

q_1 = value of the $(\frac{N}{4})^{th} item$

= value of the $(\frac{200}{4})^{th} item$

= value of the (50) *item*

This lies in cf 90 so lower quartile group is 10-15

$Q_1 = l+(q_1-c)\frac{h}{f}$

Here l= 10 , q_1 = 50, c= 40 , h= 5 and f = 50

$$Q_1 = 10+(50-40)\ \frac{5}{50} = 11$$

$$Q_1 = 11$$

Step III: Third Quartile:

q_3 = value of the $\left(3*(\frac{N}{4})^{th}\right)$ *item*

= value of the $\left(3*(\frac{200}{4})^{th}\right)$ *item*

= value of the (3* 50) *item*

= value of the (150th) *item*

This is cf 150 so lower quartile group is 15-20

$$Q_3 = l+(q_3-c)\frac{h}{f}$$

Here l = 15 , q_3 = 150, c= 90 , h= 5 and f = 60

$$Q_3 = 15+(150-90)\frac{5}{60} = 20$$

Q_3 = 20

Step IV: Eighth Decile:

d_8 = value of the $\left(8*(\frac{N}{10})^{th}\right)$ *item*

= value of the $\left(8*(\frac{200}{10})^{th}\right)$ *item*

= value of the (8* 20) *item*

= value of the (160th) *item*

This lies in cf 190 so lower quartile group is 20-25

$$D_8 = l+(d_8-c)\frac{h}{f}$$

Here l= 20 , d_8 = 190, c= 150 , h= 5 and f = 40

$$Q_1 = 20 +(160-150)\frac{5}{40} = 21.25$$

$$D_8 = 21.25$$

Step V: Ninetieth Percentile

P_{90} = value of the $\left(90*(\frac{N}{100})^{th}\right)item$

= value of the $\left(90*(\frac{200}{100})^{th}\right)item$

= value of the (180th) *item*

This lies in cf 190 so lower quartile group is 20-25

$Q_{190} = l + (q_{190} - c)\frac{h}{f}$

Here l= 20 , q_{190} = 180, c= 150 , h= 5 and f = 40

$Q_1 = 20 + (180-150)\frac{5}{40} = 23.75$

$= 23.75$

Flow Diagram:

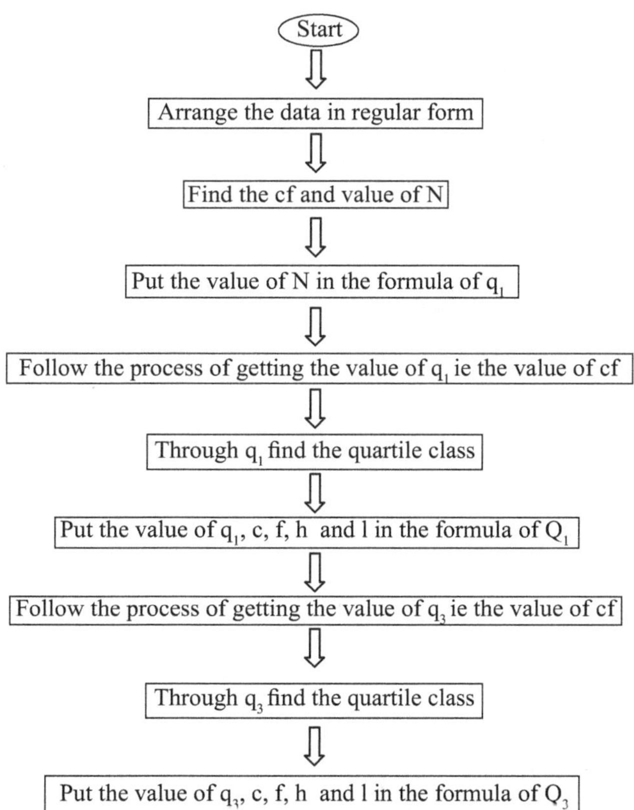

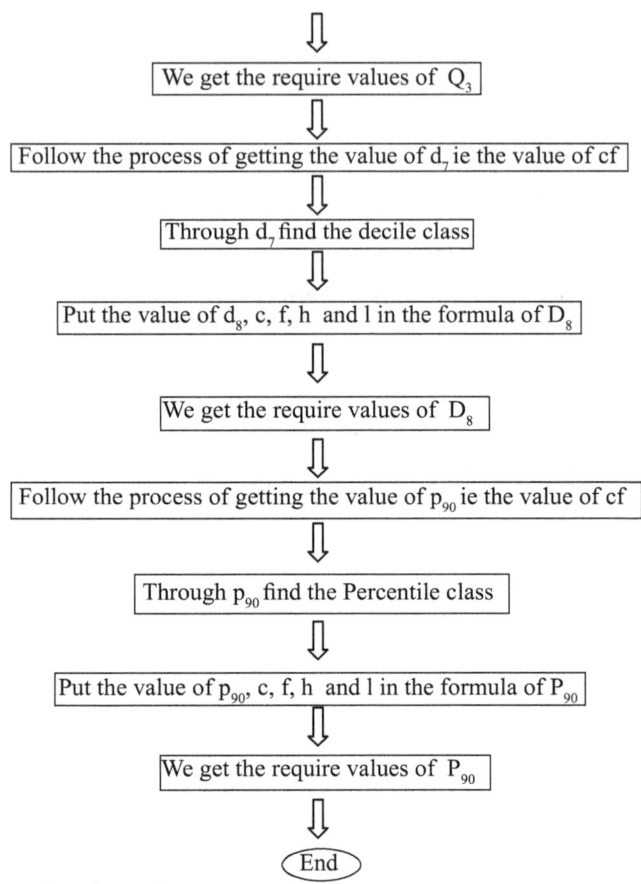

Result: From the given data we get

(i) $Q_1 = 11$

(ii) $Q_3 = 20$

(iii) $D_8 = 21.25$

(iv) $P_{90} = 23.75$

Related Questions:

1. What is partition value?
2. What is quartile?
3. What is octile?
4. What is decile?
5. What is percentile?

7

Study of Quartile Deviation

A relative measure of dispersion based on the quartile deviation is called the coefficient of quartile deviation. It is defined as

$$Q.D. = \frac{\frac{Q_3 - Q_1}{2}}{\frac{Q_3 - Q_1}{2}}$$

$$= \frac{Q_3 - Q_1}{Q_3 + Q_1}$$

It is pure number free of any units of measurement. It can be used for comparing the dispersion in two or more than two sets of data.

Objective: The wheat production (in Kg) of 20 acres is given as: 1120, 1240, 1320, 1040, 1080, 1200, 1440, 1360, 1680, 1730, 1785, 1342, 1960, 1880, 1755, 1720, 1600, 1470, 1750, and 1885. Find the quartile deviation and coefficient of quartile deviation.

Theory: quartile deviation is defined as

$$Q.D. = \frac{Q_3 - Q_1}{Q_3 + Q_1}$$

Where

Q_1 = Value of $\left(\frac{n+1}{4}\right)^{th}$ item

Q_3 = Value of $\left(\frac{3(n+1)}{4}\right)^{th}$ item

Process: Step I:

After arranging the observations in ascending order, we get 1040, 1080, 1120, 1200, 1240, 1320, 1342, 1360, 1440, 1470, 1600, 1680, 1720, 1730, 1750, 1755, 1785, 1880, 1885, 1960.

Step II:

$$Q_1 = \text{Value of } \left(\frac{n+1}{4}\right)^{th} \text{ item}$$

$$= \text{Value of } \left(\frac{20+1}{4}\right)^{th} \text{ item}$$

$$= \text{Value of } (5.25)^{th} \text{ item}$$

$$= 5^{th} \text{ item} + 0.25(6^{th} \text{ item} - 5^{th} \text{ item})$$

$$= 1240 + 0.25(1320 - 1240)$$

$$Q_1 = 1260$$

Step III:

$$Q_3 = \text{Value of } \left(\frac{3(n+1)}{4}\right)^{th} \text{ item}$$

$$= \text{Value of } \left(\frac{3(20+1)}{4}\right)^{th} \text{ item}$$

$$= \text{Value of } (15.75)^{th} \text{ item}$$

$$= 15^{th} \text{ item} + 0.75\ (16^{th} \text{ item} - 15^{th} \text{ item})$$

$$= 1750 + 0.75\ (1755 - 1750)$$

$$Q_3 = 1753.75$$

Step IV:

$$\text{Quartile Deviation (Q.D)} = \frac{Q_3 - Q_1}{2} = \frac{1753.75 - 1260}{2} = \frac{493.75}{2}$$

$$= 246.875$$

Step V:

$$\text{Coefficient of Quartile Deviation} = \frac{Q_3 - Q_1}{Q_3 + Q_1} = \frac{1753.75 - 1260}{1753.75 + 1260}$$

$$= 0.164$$

Flow Diagram

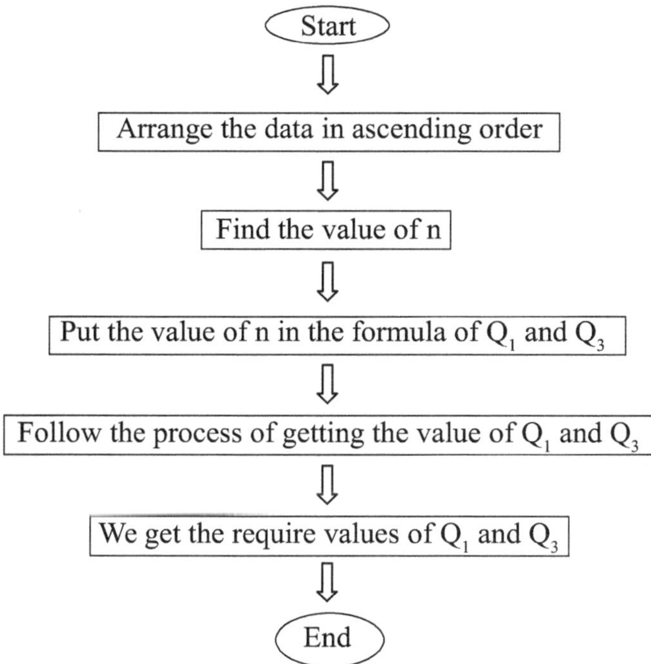

Result: The Quartile Deviation for wheat production is 246.88 and Coefficient of Quartile Deviation is 0.164.

Related Questions

Q.1 What is dispersion?

Q.2 What is the importance of dispersion?

Q.3 How many measures of dispersion?

Q.4 What is quartile deviation?

Q.5 What is coefficient of quartile deviation?

8

Study of Mean Deviation for Ungrouped Data

MEAN DEVIATION

The mean deviation is the first measure of dispersion that we will use that actually uses each data value in its computation. It is the mean of the distances between each value and the mean. It gives us an idea of how spread out from the center the set of values is.

The mean deviation is based on all the observations, a property which is not possessed by the range and the quartile deviation. The formula of the mean deviation gives a mathematical impression that is a better way of measuring the variation in the data. Any suitable average among the mean, median or mode can be used in its calculation but the value of the mean deviation is minimum if the deviations are taken from the median.

FOR UNGROUPED DATA

Objective: 5 Varieties of wheat's are tested and had scores of 92, 75, 95, 90, and 98. Find the mean deviation for wheat test.

Theory:

Mean Deviation

$$\text{MD} = \frac{\Sigma\left|X - \bar{X}\right|}{n}$$

Process: Step I: First find the mean.

$$\bar{X} = \frac{\sum x}{n}$$

$$= \frac{92+75+95+90+98}{5} = \frac{450}{5} = 90$$

Step II:

Now subtract the mean from each score, take the absolute value of each difference, total the absolute values, then divide by the number of values.

X	X - $\bar{X}$	|X - $\bar{X}$|
92	2	2
75	-15	15
95	5	5
90	0	0
98	8	8
Total		**30**

Step III:

$$\text{MD} = \frac{\sum\left|X - \bar{X}\right|}{n}$$

$$= \frac{30}{5}$$

$=6$

Step IV:

MD = 6

Flow diagram:

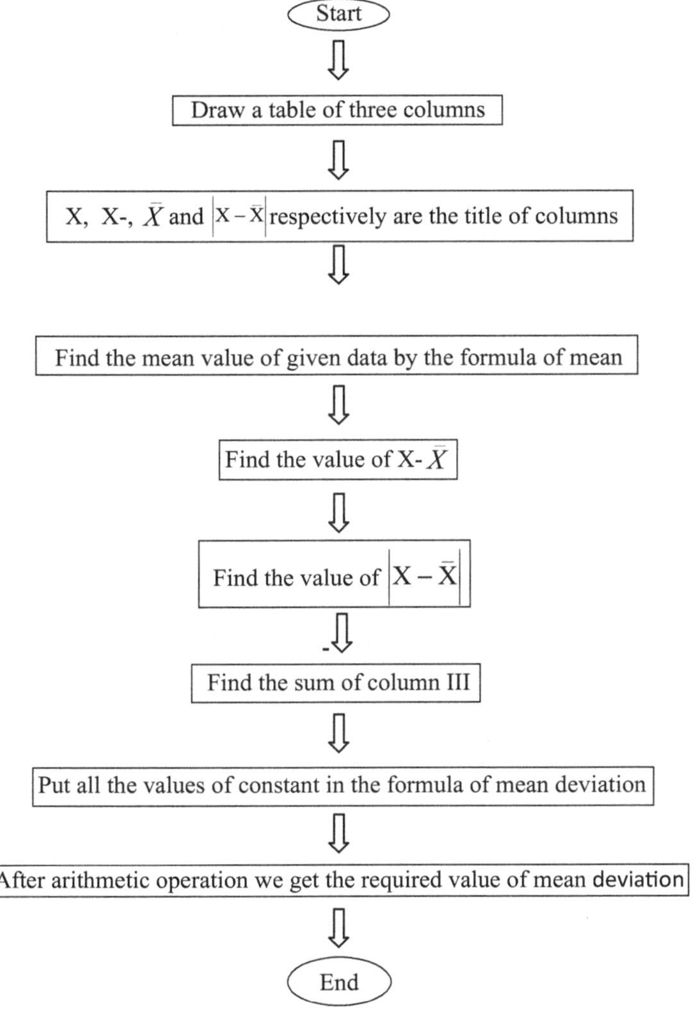

Result : We can say that on the average, wheat verities test scores deviated by 6 points from the mean.

Related Question

Q.1 What is mean deviation?

Q.2 Why modulus used in mean deviation?

Q.3 What is mean deviation from mean?

Q.4 What is mean deviation from median?

Q.5 Which is mean deviation from mode?

9

Study of Mean Deviation for Grouped Data

COEFFICIENT OF THE MEAN DEVIATION

A relative measure of dispersion based on the mean deviation is called the coefficient of the mean deviation or the coefficient of dispersion. It is defined as the ratio of the mean deviation to the average used in the calculation of the mean deviation.

FOR GROUPED DATA

Objective: Calculate the mean deviation from mean and its coefficients from the following data.

Size of Items	3-4	4-5	5-6	6-7	7-8	8-9	9-10
Frequency	3	7	22	60	85	32	8

Theory: For frequency distribution, the mean deviation is given by

$$MD = \frac{\Sigma f\left|X - \bar{X}\right|}{\Sigma f}$$

$$\text{Coefficient of MD (about mean)} = \frac{\text{Mean Deviation from Mean}}{\text{Mean}}$$

Process: Step I:

Size of Items	**X**	**f**	**fX**	$\mathbf{\left\|X-\bar{X}\right\|}$	$\mathbf{f\left\|X-\bar{X}\right\|}$
3-4	3.5	3	10.5	3.59	10.77
4-5	4.5	7	31.5	2.59	18.13
5-6	5.5	22	121	1.59	34.98
6-7	6.5	60	390	0.59	35.40
7-8	7.5	85	637.5	0.41	34.85
8-9	8.5	32	272	1.41	45.12
9-10	9.5	8	76	2.41	19.28
Total		**217**	**1538.5**		**198.53**

Step II:

Mean $\bar{X} = \frac{\sum fX}{\sum f} = \frac{1538.5}{217}$

$= 7.09$

Step III:

MD from Mean $= \frac{\sum f\left|X - \bar{X}\right|}{\sum f} = \frac{198.53}{217}$

$= 0.915$

Step IV:

Coefficient of MD (Mean) $= \frac{\text{MD from mean}}{\text{mean}}$

$= \frac{0.915}{7.09}$

$= 0.129$

Flow Diagram

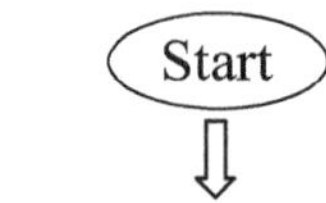

Draw a table of six columns

Size of items, X, f, Xf, $\left|X - \bar{X}\right|$ and f $\left|X - \bar{X}\right|$ are the title of columns respectively

Write down the value of size of items and f according to the given objective

Find the value of X by mid value of age

Find the value of fx by multiplying second column to third column

Find the value of mean by the formula

Find the value of $\left|X - \bar{X}\right|$ by subtracting the values of x fro mean

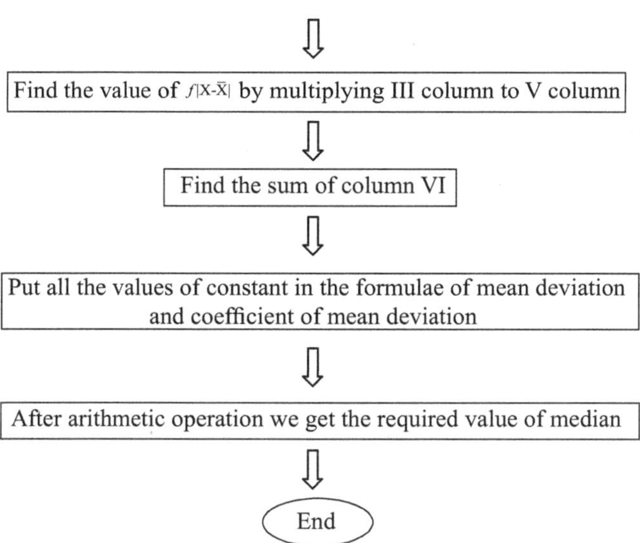

Result: From the given data

(i) The M.D. for the given data is 0.195.

(ii) The Coefficient of M.D. is 0.129.

Related Questions

Q.1 What are the characteristics of mean deviation?

Q.2 What is coefficient of mean deviation?

Q.3 Is mean deviation always same?

Q.4 Is mean deviation is negative?

Q.5 Why mean deviation is useful?

10

Study of Standard Deviation for Ungrouped Data

STANDARD DEVIATION AND THEIR COEFFICIENT

The standard deviation is a measure of the spread of scores within a set of data. Usually, we are interested in the standard deviation of a population. However, as we are often presented with data from a sample only, we can estimate the population standard deviation from a sample standard deviation. These two standard deviations - sample and population standard deviations - are calculated differently. In statistics, we are usually presented with having to calculate sample standard deviations.

"Square root of the mean of the square of the sum of deviation taken from their mean"

FOR UNGROUPED DATA

Objective: Calculate the coefficient of standard deviation and coefficient of standard deviation for the following sample data: 120, 110, 115, 122, 126, 140, 125, 121, 120 and 131.

Theory: The population standard deviation formula is:

$$\sigma = \sqrt{\frac{\Sigma\left(x-\bar{x}\right)^2}{n}}$$

where,

σ = population standard deviation

Σ= sum of

$\bar{x}$ = population mean

n = number of scores in sample.

Coefficient of Standard Deviation = $\frac{\sigma}{\bar{X}}$

Process: Step I

X	$(X-\bar{X})$	$(X-\bar{X})^2$
120	-3	9
110	-13	169
115	-8	64
122	-1	1
126	3	9
140	17	289
125	2	4
121	-2	4
120	-3	9
131	8	64
1230		**622**

Step II

Arithmetic Mean $\mu = \bar{X} = \frac{\sum X_i}{n}$

$= 1230/10$

$= 123$

Step III

Then Standard Deviation $\sigma = \sqrt{\frac{622}{10}}$

$= \sqrt{62.2}$

$= 7.89$

Step IV

Coefficient of Standard Deviation $= \frac{\sigma}{\bar{X}}$

$=7.89/123$

$= 0.064.$

Flow Diagram

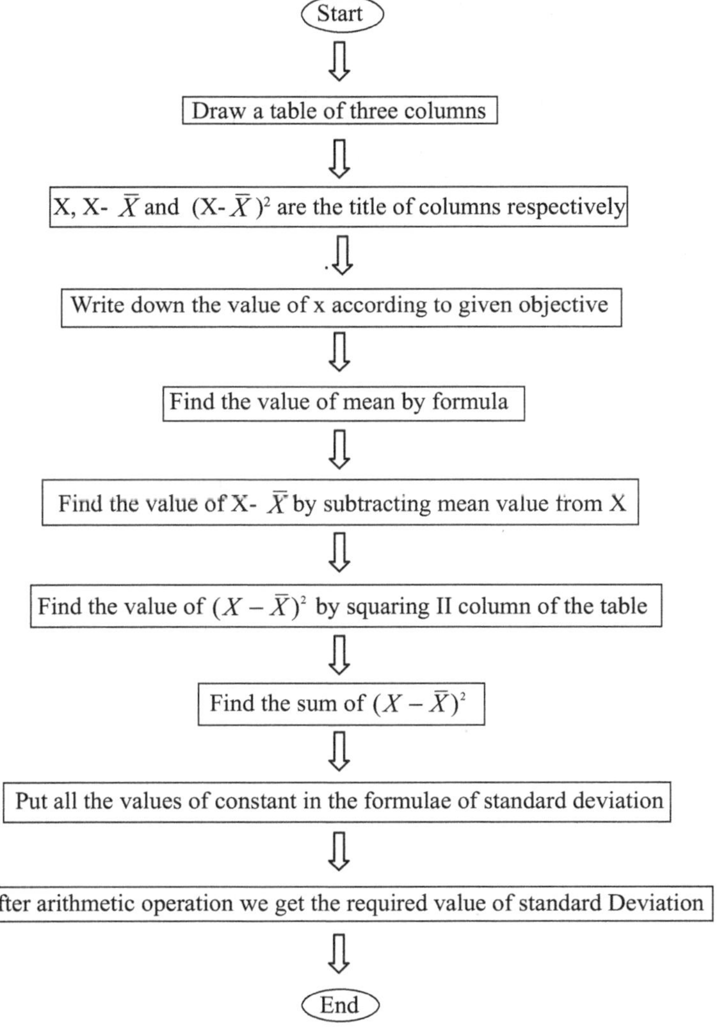

Result: From the given data we get

(i) Standard Deviation = 7.89

(ii) Coefficient of Standard Deviation = 0.064.

Related Questions

Q.1 What is standard deviation?

Q.2 What is variance?

Q.3 Why it is best measures of dispersion?

Q.4 What is coefficient of standard deviation?

Q.5 What is the symbol of standard deviation?

11

Study of Standard Deviation for Grouped Data

FOR GROUPED DATA

Objective: The following data shows the marks of students and their frequencies find the standard deviation and its coefficient.

Marks:	1-3	3-5	5-7	7-9
Frequency:	40	30	20	10

Theory:

Mean $\bar{X} = \frac{\sum fX}{\sum f}$

Standard Deviation $\sigma = \sqrt{\frac{f(X-\bar{X})^2}{\sum f}}$

Coefficient of Standard deviation $= \frac{\sigma}{\bar{X}}$

Process:

Step I: Table

Marks	f	**X**	f**X**	$(X-\bar{X})^2$	$f(X-\bar{X})^2$
1-3	40	2	80	4	160
3-5	30	4	120	0	0
5-7	20	6	120	4	80
7-9	10	8	80	16	160
Total	**100**		**400**		**400**

Step II:

$$\bar{X} = \frac{\sum fX}{\sum f} = \frac{400}{100}$$

$$= 4$$

Step III:

$$\sigma = \sqrt{\frac{f(X-\bar{X})^2}{\Sigma f}} = \sqrt{\frac{400}{100}}$$

= 2 Marks

Step IV:

$$\text{Coefficient of Standard Deviation} = \frac{\sigma}{\bar{X}}$$

$$= \frac{2}{4}$$

$$= 0.5$$

Flow Diagram:

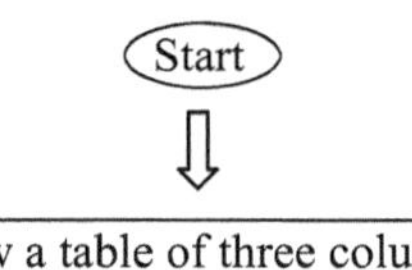

Draw a table of three columns

Marks, f, X, fX , X- $\bar{X}$ and f (X-$\bar{X}$)^2 are the title of columns respectively

Write down the value of marks and f according to given objective

Find the values of x by mid value of marks

Multiplying II and III column and get the value of fx

Find the value of mean by formula

Find the value of X- $\bar{X}$ by subtracting mean value from X

Find the value of $(X-\bar{X})^2$ by squaring II column of the table

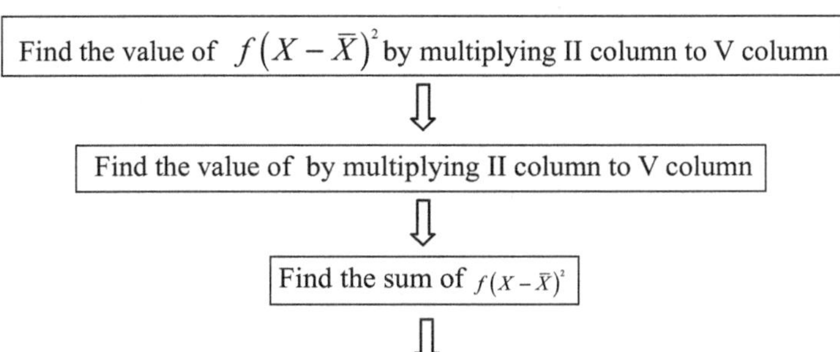

Put all the values of constant in the formulae of standard deviation and coefficient of standard deviation.

After arithmetic operation we get the required value of standard Deviation, coefficient of SD

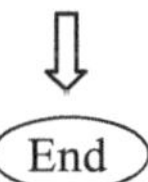

Result : From the given data

(i) Standard Coefficient = 2 Marks

(ii) Coefficient of Standard Deviation = 0.5

(iii) Coefficient of variation = 50%

12

Study of Coefficient of Variation for Ungrouped Data

COEFFICIENT OF VARIATION:

The most important of all the relative measure of dispersion is the coefficient of variation. This word is variation not variance. There is no such thing as coefficient of variance. The coefficient of variation(C.V.) is defined as: $\frac{\sigma}{\overline{X}}*100$

Thus C.V. is the value of σ when $\overline{X}$ is assumed equal to 100. It is a pure number and the unit of observations is not mentioned with its value. It is written in percentage form like 20% or 25%. When its value is 20%, it means that when the mean of the observations is assumed equal to 100, their standard deviation will be 20. The C.V. is used to compare the dispersion in different sets of data particularly the data which differ in their means or differ in the units of measurement. The wages of workers may be in rupees and the consumption of meat in their families may be in kilograms. The standard deviation of wages in rupees cannot be compared with the standard deviation of amounts of meat in kilograms. Both the standard deviations need to be converted into coefficient of variation for comparison. Suppose the value of C.V. for wages is 10% and the values of C.V. for kilograms of meat is 25%. This means that the wages of workers are consistent. Their wages are close to the overall average of their wages. But the families consume meat in quite different quantities. Some families use very small quantities of meat and some others use large quantities of meat. We say that there is greater variation in their consumption of meat. The observations about the quantity of meat are more dispersed or more variant.

FOR UNGROUPED DATA

Objective: Price of Rice in five years in two cities A and B are Given below, find the city which had more stable price of rice:

City A:	20	22	19	23	16
City B:	10	20	18	12	15.

Theory: Here we compare the coefficient of variation for the two cities A and B, so we compute

$$\bar{X} = \frac{\Sigma X_i}{n} \text{ and } \sigma = \sqrt{\frac{(X-\bar{X})^2}{n}}$$

Coefficient of variation for two cities A and B

$$(\text{C.V.})_A = \frac{\sigma}{\bar{X}} * 100$$

$$(\text{C.V.})_B = \frac{\sigma}{\bar{X}} * 100$$

If $(\text{C.V.})_A < (\text{C.V.})_B$ then A is more consistent than B.

Process: Step I:

Coefficient of variation for city A:

X	d = X-$\bar{X}$	d²
20	0	0
22	+2	4
19	-1	1
23	+3	9
16	-4	16
100		**30**

Step II:

$$\bar{X} = \frac{\Sigma X_i}{n} \text{ and } \sigma = \sqrt{\frac{(X-\bar{X})^2}{n}}$$

$= \frac{100}{5}$ $\qquad \sigma = \sqrt{\frac{30}{100}}$

$= 20$ $\qquad = \sqrt{6}$

$= 2.45$

Step III:

$$\text{C.V.} = \frac{\sigma}{\bar{X}} * 100$$

$$= \frac{2.45}{20} * 100$$

$= 12.25\%$

Coefficient of variation for city B:

Step I:

X	d = X- $\bar{X}$	d²
10	-5	25
20	+5	25
18	+3	9
12	-3	9
15	0	0
75		**68**

Step II:

$\bar{X} = \frac{\Sigma X_i}{n}$ and $\sigma = \sqrt{\frac{(X-\bar{X})^2}{n}}$

$= \frac{75}{5}$ $\quad\quad$ $= \sqrt{\frac{68}{5}}$

$= 15$ $\quad\quad$ $= \sqrt{13.6}$

$= 3.6$

Step III:

$C.V. = \frac{\sigma}{\bar{X}} * 100$

$= \frac{3.69}{20} * 100$

$= 24.6\%$

Since $(C.V.)_A < (C.V.)_B$ the city A has more stable price.

Flow Diagram:

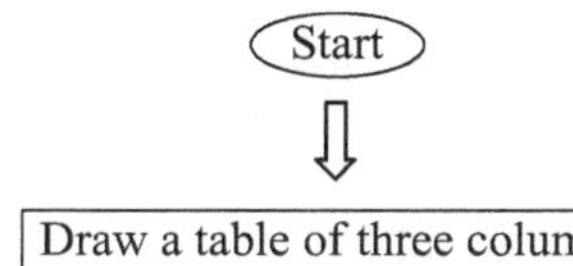

Draw a table of three columns

X, d= X- $\bar{X}$ and d² = $(X-\bar{X})^2$ are the title of columns respectively

Write down the value of x for city A according to given objective

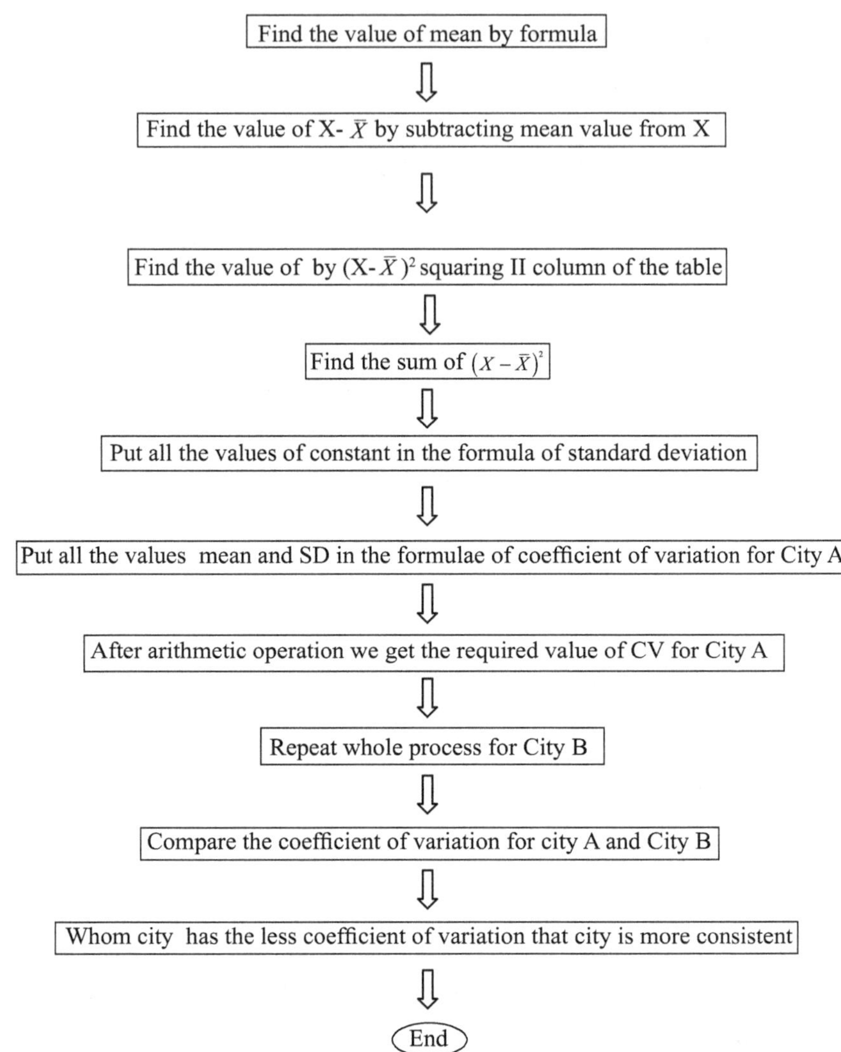

Result: Hence city A has more stable price than city B.

Related Question

Q.1 What is coefficient of variation?

Q.2 Why it is useful?

Q.3 When coefficient of variation essential?

Q.4 What is consistency?

Q.5 Why coefficient of variation used in percentage?

13

Study of Coefficient of Variation for Grouped Data

FOR GROUPED DATA

Objective: Data regarding lives of two new models of refrigerators collected in a recent survey are

Life (in years):	0-2	2-4	4-6	6-8	8-10	10-12.
Model A:	5	16	13	7	5	4
Model B:	2	7	12	19	9	1

What is the average life of each model of these refrigerators? Which model has greater uniformity?

Theory:

Arithmetic Mean: $\overline{X} = a + \frac{\Sigma fd}{\Sigma f} * i$

Standard Deviation $\sigma = \sqrt{\frac{\Sigma fd^2}{\Sigma f} - \left(\frac{\Sigma fd}{\Sigma f}\right)^2} * i$

where

a = Assumed Mean

d = $\frac{X - a}{h}$

i = Class Interval

Process: For model A

Step I: Here a =5 i= 2

Class Interval	Frequency	Mid Point	$d = \frac{X - a}{h}$	fd	fd^2
0-2	5	1	-2	-10	20
2-4	16	3	-1	-16	16
4-6	13	5	0	0	0
6-8	7	7	1	7	7
8-10	5	9	2	10	20
10-12	4	11	3	12	36
Total	**50**			**3**	**99**

Step II:

Arithmetic Mean $\overline{X} = 5 + \frac{3}{50} * 2$

$= 5.12$ year

Step III:

Standard Deviation $\sigma = \sqrt{\frac{\sum fd^2}{\sum f} - \left(\frac{\sum fd}{\sum f}\right)^2} * i$

$= \sqrt{\frac{99}{50} - \left(\frac{3}{50}\right)^2} * 2$

$= \sqrt{1.9764} * 2$

$= 1.406 * 2$

$= 2.812$ year

Step IV:

Coefficient of variation for model A $= \frac{\sigma}{\overline{X}} * 100\%$

$= \frac{2.812}{5.12} * 100\%$

$= 54.92\%$

For model B:

Step I: Here a =5 i= 2

Class Interval	Frequency	Mid Point	$d = \frac{X-a}{h}$	fd	fd^2
0-2	2	1	-2	-4	8
2-4	7	3	-1	-7	7
4-6	12	5	0	0	0
6-8	19	7	1	19	19
8-10	9	9	2	18	36
10-12	1	11	3	3	9
Total	**50**			**29**	**79**

Step II:

Arithmetic Mean $\bar{X} = 5 + \frac{29}{50} * 2$

$= 6.16$ year

Step III:

Standard Devaiation $\sigma = \sqrt{\frac{\Sigma fd^2}{\Sigma f} - \left(\frac{\Sigma fd}{\Sigma f}\right)^2} * i$

$= \sqrt{\frac{79}{50} - \left(\frac{29}{50}\right)^2} * 2$

$= \sqrt{1.2436} * 2$

$= 1.115 * 2$

$= 2.230$ year

Step IV:

Coefficient of variation for model B $= \frac{\sigma}{\bar{X}} * 100\%$

$= \frac{2.23}{6.16} * 100\%$

$= 36.20\%$

Step V: Since $(C.V.)_A > (C.V.)_B$, Model B is more uniform.

Flow Diagram:

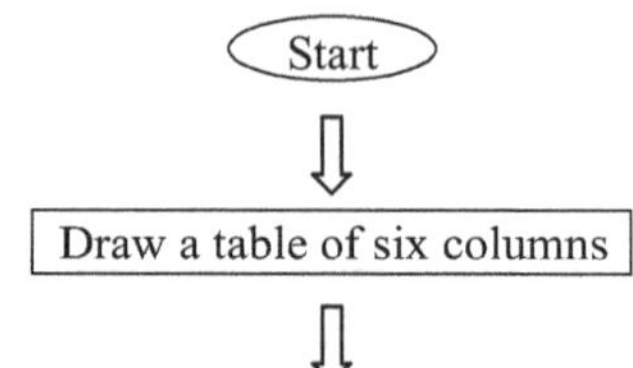

Class interval, f, x, d, fd and fd^2 are the title of columns respectively

Write down the value of class interval and f according to the given objective

Find the value of x by mid value of class interval

Find the value of d by $\frac{X-a}{h}$

Find the value of fd by multiplying second column to fourth column

Find the value of mean by the formula

Find the value of standard deviation by the formula

Put all the values of constant in the formulae of coefficient of variation for model A

After arithmetic operation we get the required value of CV for model A

Repeat whole process for model B

Compare the coefficient of variation for model A and model B

Whom model has the less coefficient of variation that model is more consistent

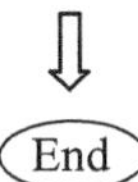

End

Results: From the given data we get

(1) The average life of model A is 5.12 year.

(2) The average life of model B is 6.16 year.

(3) Model B is more uniform than model A.

Related Question:

1. How CV effect the consistency?
2. What is standard error?
3. What is d?
4. What is I?
5. What is a?

14

Study of Moments for Ungrouped and Grouped Data

MOMENTS

The term moment in mechanics refers to the measure of a force with reference to its tendency to produce rotation. The strength of this tendency is dependent upon the amount of the force and the distance from the origin at which the force is applied. If on both the sides of the origin the force are equal than there will be a balance, at a balanced position the positive product equals the negative product.

Moment is used in Statistics in a quite analogous sense. The class frequencies are looked upon as the forces and the deviations of the different values from the mean are taken as the distances. In other words, moment is the mean of the first, second, third, fourth, etc. powers of deviations of the different values from the arithmetic mean. The first moment about the mean is always zero and second moment about the mean is variance.

If the moment calculated from arithmetic mean are called central moments and denoted by μ with a subscript which indicate the order of the moment. Moments are the mean values of powers of the deviations in any frequency distribution taken about three points

(i) Origin

(ii) Mean

(iii) Any other point

Moment about origin

In case of individual: The rth order moment about origin is denoted by and is defined as

$$\mu\,\mathrm{r} = \mu_r = \frac{\sum x^r}{n}\ 1,2,3,4\ldots$$

In case of frequency distribution series:

$$\mu_r = \frac{\Sigma f x^r}{\Sigma f} \quad 1,2,3,4\ldots$$

Moment about Mean:

In case of individual:

$$\mu_r = \frac{\Sigma(x-\bar{x})^r}{n} \quad r = 1,2,3,4\ldots \text{ where } \bar{x} = \frac{\Sigma x_i}{n}$$

In case of frequency distribution series:

$$\mu_r = \frac{\Sigma f(x-\bar{x})^r}{\Sigma f} \quad r = 1,2,3,4\ldots$$

Moment about any point:

In case of individual:

$$\mu_r = \frac{\Sigma(x-a)^r}{n} \quad r = 1,2,3,4\ldots \text{ where}$$

In case of frequency distribution series:

$$\mu_r = \frac{\Sigma f(x-a)^r}{\Sigma f} \quad r = 1,2,3,4\ldots$$

Relationship between central moments and the moments about any arbitrary point:

$$\mu_1 = 0$$

$$\mu_2 = \mu_2^{'} - \left(\mu_1^{'}\right)^2$$

$$\mu_3 = \mu_3^{'} - 3\mu_1^{'}\mu_2^{'} + 2\left(\mu_1^{'}\right)^3$$

$$\mu_4 = \mu_4^{'} - 4\mu_1^{'}\mu_3^{'} + 6\left(\mu_1^{'}\right)^2\mu_2^{'} - 3\left(\mu_1^{'}\right)^4$$

Objective (a): Compute the first four central moments of the data

12 14 16 18 20.

Theory: The central moment is given by the following formula

$$\mu_r = \frac{\Sigma(x-\bar{x})^r}{n} \quad r = 1,2,3,4\ldots \text{ where}$$

So that

$\mu_1 = 0$

$$\mu_2 = \frac{\sum(x-\overline{x})^2}{n}$$

$$\mu_3 = \frac{\sum(x-\overline{x})^3}{n}$$

and

$$\mu_4 = \frac{\sum(x-\overline{x})^4}{n}$$

Process : Step I

X	x-$\overline{x}$	(x-$\overline{x}$)^2	(x-$\overline{x}$)^3	(x-$\overline{x}$)^4
12	-4	16	-64	256
14	-2	4	-8	16
16	0	0	0	0
18	2	4	8	16
20	4	16	64	256
80	**0**	**40**	**0**	**544**

Step II

$$\overline{x} = \frac{\Sigma x_i}{n} = \frac{80}{5} = 16$$

Step III

$\mu_1 = 0$

$$\mu_2 = \frac{\sum(x-\overline{x})^2}{n}$$

$$\mu_2 = \frac{40}{5} = 8$$

$$\mu_3 = \frac{\sum(x-\overline{x})^3}{n}$$

$\mu_3 = 0$

$$\mu_4 = \frac{\sum(x-\overline{x})^4}{n}$$

$$\mu_4 = \frac{544}{5} = 108.8$$

Flow Diagram:

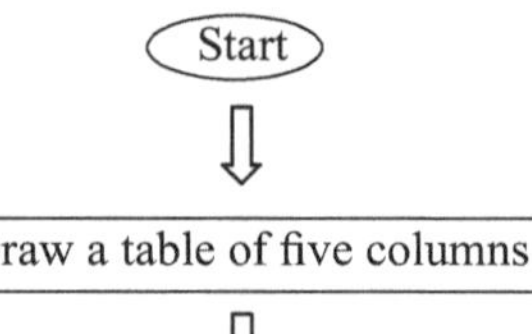

Marks, X, X-$\overline{X}$, $(X-\overline{X})^2$, $(X-\overline{X})^3$ and $(X-\overline{X})^4$ are the title of columns respectively

Write down the value of x according to given objective

Find the value of mean by formula

Find the value of X-$\overline{X}$ by subtracting mean value from $(X-\overline{X})^2$

Find the value of $(X-\overline{X})^3$ by squaring II column of the table

Find the value of $(X-\overline{X})^4$

Find the total of all the columns

Put all the values of constant in the formulae of central momnts.

After arithmetic operation we get the required value of central moments.

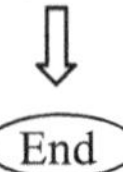

End

Result: The values of central moments are

(i) $\mu_1 = 0$ (ii) $\mu_2 = 8$ (iii) $\mu_3 = 0$ (iv) $\mu_4 = 108.8$

Objective (b): Calculate first three moments about mean 89 for the following table

Marks:	81	87	89	90	91	94	96
No. of Students :	7	11	15	8	4	3	2

Theory: The moment about any arbitrary point is

$$\mu'_r = \frac{\Sigma(x-a)^r}{n} \quad r = 1,2,3,4\ldots$$

where a is the arbitrary point.

Process: Step I: here a=89

x	f	d = (x-89)	fd	fd²	fd³
81	7	-8	-56	448	-3584
87	11	-2	-22	44	-88
89	15	0	0	0	0
90	8	1	8	8	8
91	4	2	8	16	32
94	3	5	15	75	375
96	2	7	14	98	686
Total	**50**		**-33**	**689**	**-2571**

Step II:

The first moment about 89 is

$$\mu'_1 = \frac{\Sigma f(x-89)}{50} = \frac{-33}{50} = \text{-0.66 marks}$$

The second moment about 89 is

$$\mu'_2 = \frac{\Sigma fd^2}{50} = \frac{689}{50} = 13.78 \text{ marks}$$

The third moment about 89 is

$$\mu'_3 = \frac{\Sigma fd^2}{50} = \frac{-2571}{50} = \text{-51.42 marks}$$

Flow Diagram:

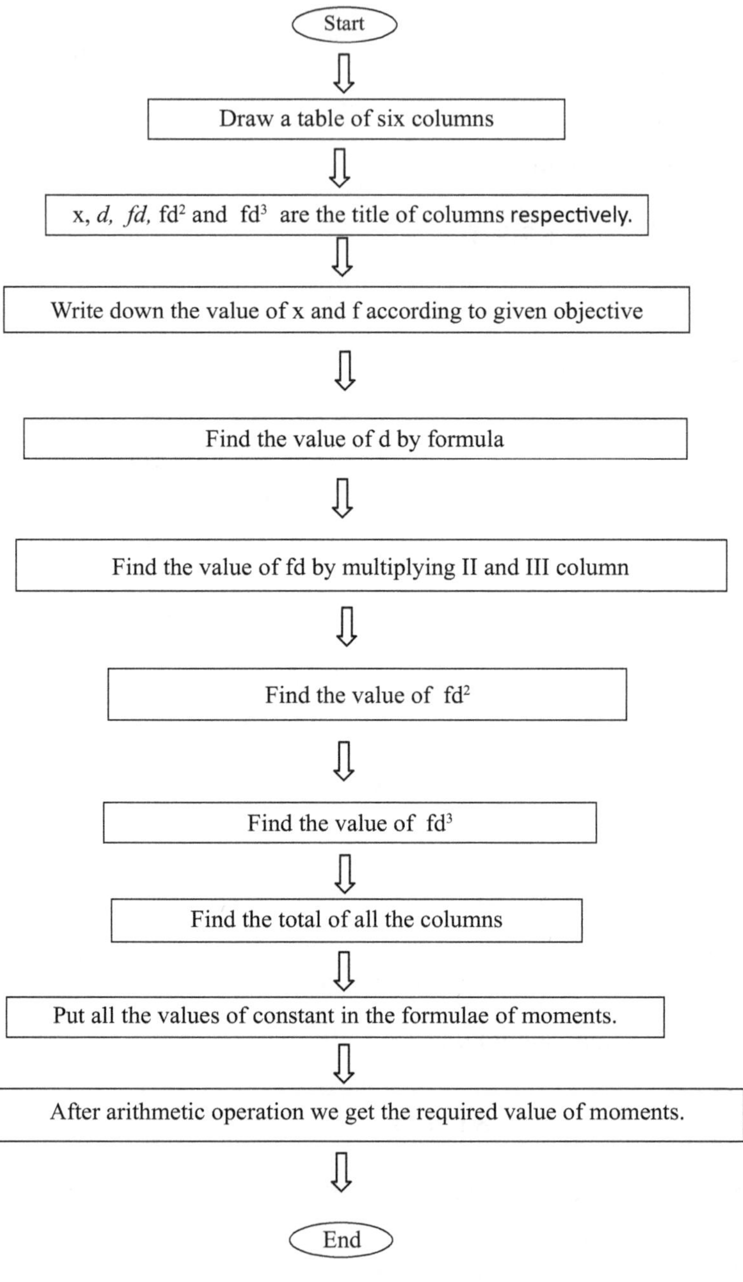

Result: The values of moments about arbitrary point are

(i) $\mu_1 = 0.66$ marks (ii) $\mu_2 = 13.78$ marks (iii) $\mu_3 = 51.42$ marks

Related Questions

Q.1 What is moments?

Q.2 What is central moments?

Q.3 What is raw moments?

Q.4 Why first order moment about origin is zero?

Q.5 What is variance in moments?

15

Study of Skewness for Ungrouped and Grouped Data

SKEWNESS

Skewness means lack of symmetry. Skewness denotes the tendency of a distribution to depart from symmetry. A frequency distribution is said to be skewed if the frequencies decrease with markedly greater rapidity on one side of the central maximum than on the other side. This characteristic of a frequency distribution is known as skewness and the measures of asymmetry are usually called measures of skewness.

The main object of measuring skewness is to know the direction of the variation from an average and to compare the frequency distribution and the shape of their curves.

There are two types of skewness

(i) Positive Skewness (ii) Negative Skewness

(i) Positive Skewness: When in a frequency distribution the frequencies are piled up at the lower values of the variable and spread out over a greater range of values on the high value end, the skewness is known as positive.

(ii) Negative Skewness: When in a frequency distribution the frequencies are piled up at the higher values of the variable and spread out over a greater range of values on the low value end, the skewness is known as negative.

Measures of Skewness

Measures of skewness may be subjective or relative. Absolute measures tell us the direction and extent of symmetry in a frequency distribution.

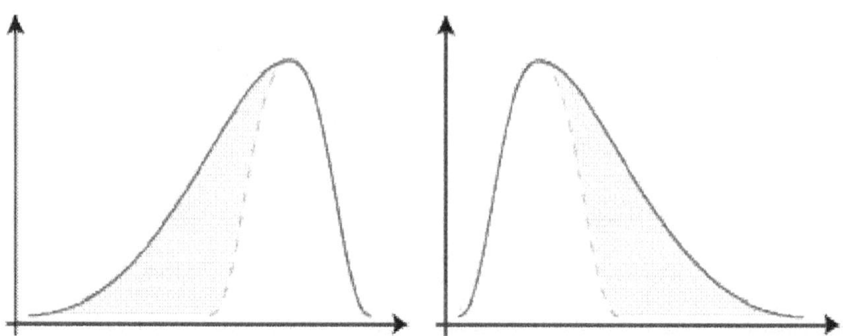

Relative measures often called coefficient of skewness , permit us to compare two or more frequency distribution.

1. **Pearson's first skewness coefficient (mode skewness)**

   The Pearson mode skewness, or first skewness coefficient, is defined as (mean − mode)/ standard deviation

2. **Pearson's second skewness coefficient (median skewness)**

The Pearson median skewness, or second skewness coefficient is defined as 3(mean − median)/ standard deviation

3. **Quantile-based measures**

   Bowley's measure of skewness also called Yule's coefficient is defined as:

$$\frac{Q_3 + Q_1 - 2Q_2}{Q_3 - Q_1}$$

4. **Moments based measures**

$$\beta_1 = \frac{\mu_3^2}{\mu_2^3}$$

UNGROUPED DATA

Objective: Find out Karl Pearson's coefficient of Skewness from the following data

Size of the Item:	7.4	8.4	9.4	10.4	11.4	12.4	13.4
Frequency:	2	6	20	14	8	6	4

Theory: Arithmetic Mean

$$\bar{X} = A + \frac{\sum fd_x}{N}$$

Mode : Highest frequency

Standard Deviation

$$\sigma = \sqrt{\frac{\sum fd_x^2}{N} - (\frac{\sum fd_x}{N})^2}$$

Karl Pearson's Coefficient of Skewnes: (mean − mode)/ standard deviation

Process: Step I: A= 10.4

Size	f	d_x = X- A	f d_x	f d_x^2
7.4	2	-3	-6	18
8.4	6	-2	-12	24
9.4	20	-1	-20	20
10.4	14	0	0	0
11.4	8	1	8	8
12.4	6	2	12	24
13.4	4	3	12	36
Total	**60**		**-6**	**130**

Step II:

Arithmetic Mean

$$\bar{X} = A + \frac{\sum fd_x}{N}$$

= 10.4 + (-6/160)

= 10.4-0.1

= 10.3

Step III: Mode =9.4

Step IV: Standard deviation

$$\sigma = \sqrt{\frac{\sum fd_x^2}{N} - \left(\frac{\sum fd_x}{N}\right)^2}$$

$$= \sqrt{\frac{130}{60} - \left(\frac{-6}{60}\right)^2}$$

$$= \sqrt{2.17 - .01}$$

$$= \sqrt{2.16}$$

=1.47

Step V:

Karl Pearsons coefficient of Skewness = (mean − mode)/ standard deviation

= (10.3-9.4)/1.47

= 0.9/1.47

= 0.61

Flow Diagram:

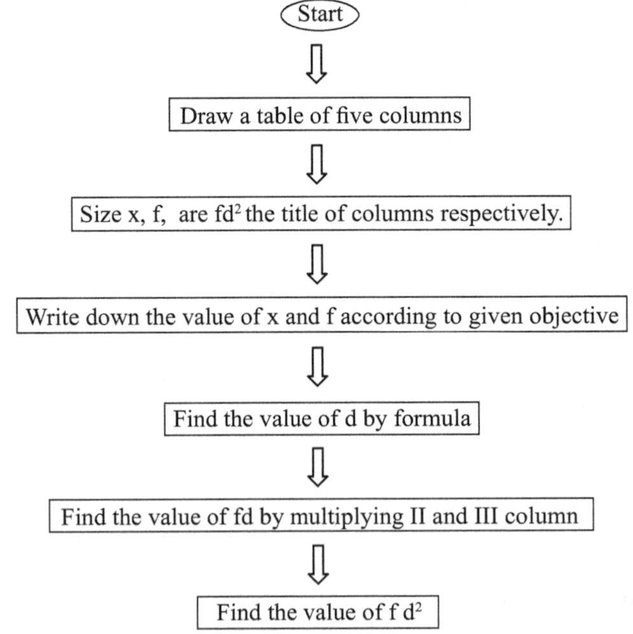

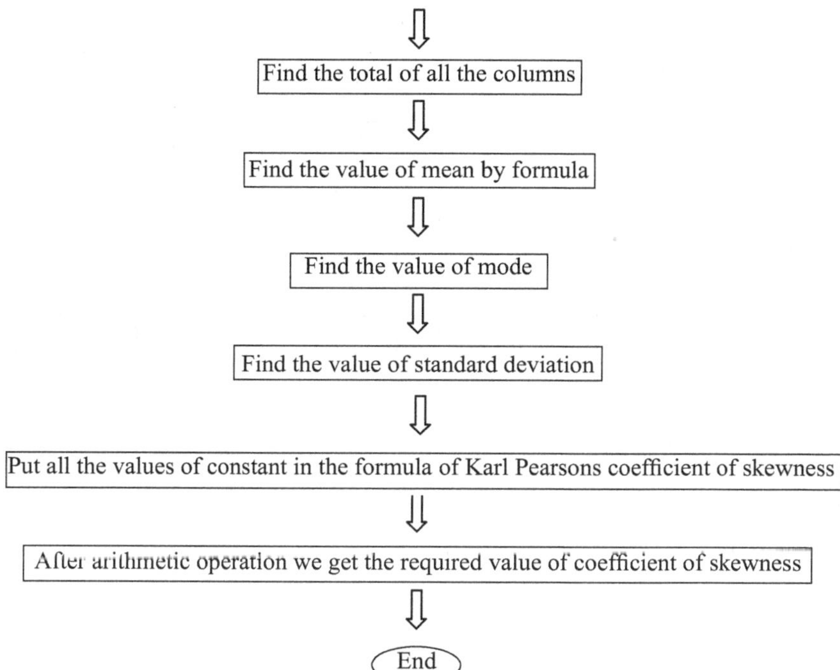

FOR GROUPED DATA

Objective:

The following are the 10 Soybean plant height collected from a particular plot

Plant height (in cm)	0-10	10-20	20-30	30-40	40-50
No. of Plants	2	2	3	2	1

Calculate coefficient of skewness based on moments.

Theory: Coefficient of skewness based on moments is

$$\beta_1 = \frac{\mu_3^2}{\mu_2^3}$$

Where μ_3 and μ_4 are the third and fourth central moments.

$$\mu_2 = \frac{\Sigma f_i(x-\bar{x})^2}{N} = \frac{\Sigma f_i(d)^2}{N}$$

$$\mu_3 = \frac{\Sigma f_i(x-\bar{x})^3}{N} = \frac{\Sigma f_i(d)^3}{N}$$

Process: Step I

Class Interval	Mid Value x	f	fx	d= x-$\overline{x}$	fd	fd^2	fd^3
0-10	5	2	10	-18	-36	648	-11664
10-20	15	2	30	-8	-16	128	-1024
20-30	25	3	75	2	6	12	24
30-40	35	2	70	12	24	288	3456
40-50	45	1	45	22	22	484	10648
		10	**230**		**0**	**1560**	**1440**

Step II

$$\overline{x} = \frac{\sum fx}{N}; N = \sum f$$

Step III

$$\mu_2 = \frac{\sum f_i (d)^2}{N}$$

$$\mu_2 = \frac{1560}{10} = 156.0$$

$$\mu_3 = \frac{\sum f_i (d)^3}{N}$$

$$\mu_3 = \frac{1440}{10} = 144.0$$

Step IV

$$\beta_1 = \frac{\mu_3^2}{\mu_2^3}$$

$$\beta_1 = \frac{(144)^2}{(156)^3} = \frac{20736}{3796416}$$

$= 0.0055$

Flow Diagram:

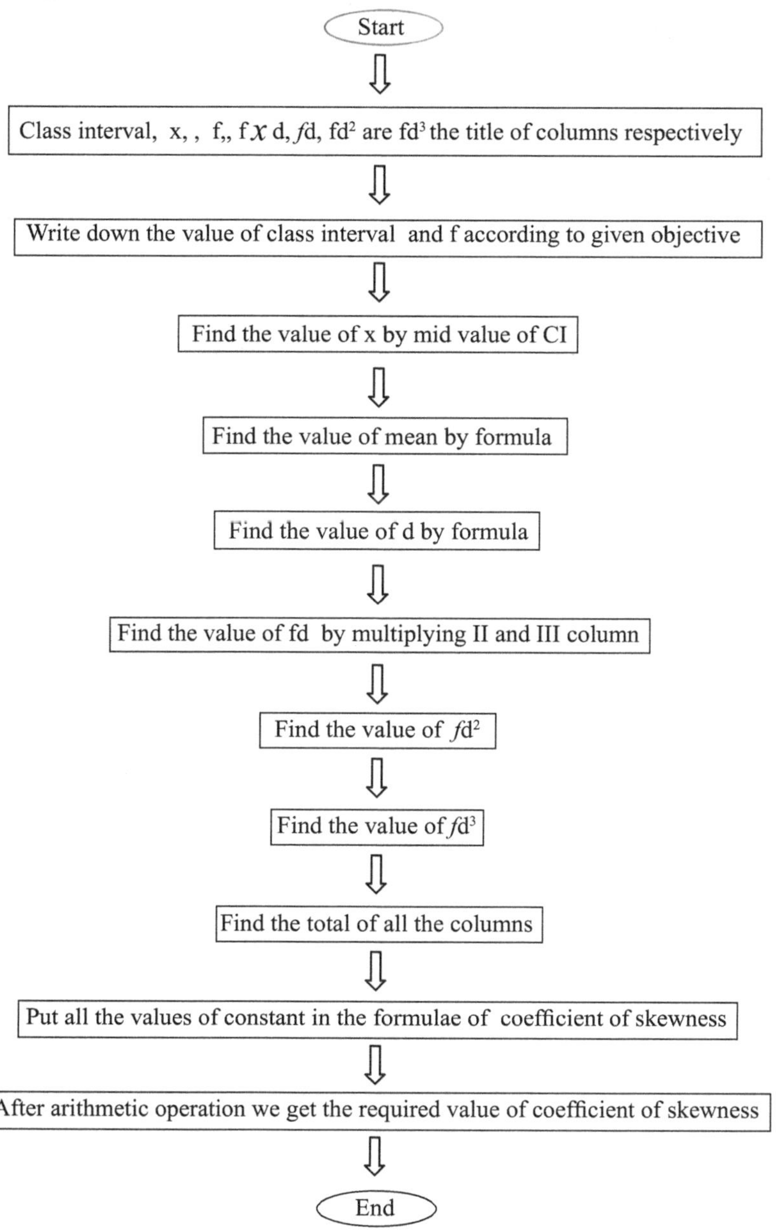

Result: From the given data the coefficient of skewness is = 0.0055

Related Questions

Q.1 What is skewness?

Q.2 What is coefficient of skewness?

Q.3 What is measures of skewness?

Q.4 What is positive skewness?

Q.5 What is negative skewness?

16

Study of Kurtosis for Ungrouped and Grouped Data

KURTOSIS

Kurtosis refers to the extent to which unimodal frequency curve is peaked. Kurtosis is a measure that refers to the peakedness of the top of the curve. Kurtosis gives the degree of flatness in the region about the mode of a frequency distribution.

Types of Kurtosis

There are three categories of kurtosis that can be displayed by a set of data. All measures of kurtosis are compared against a standard normal distribution, or bell curve.

The first category of kurtosis is a mesokurtic distribution. This type of kurtosis is the most similar to a standard normal distribution in that it also resembles a bell curve. However, a graph that is mesokurtic has fatter tails than a standard normal distribution and has a slightly lower peak. This type of kurtosis is considered normally distributed but is not a standard normal distribution.

The second category is a leptokurtic distribution. Any distribution that is leptokurtic displays greater kurtosis than a mesokurtic distribution. Characteristic of this type of distribution is one with extremely thick tails and a very thin and tall peak. The prefix of "lepto-" means "skinny," making the shape of a leptokurtic distribution easier to remember. T-distributions are leptokurtic.

The final type of distribution is a platykurtic distribution. These types of distributions have slender tails and a peak that's smaller than a mesokurtic distribution. The prefix of "platy-" means "broad," and it is meant to describe a short and broad-looking peak. Uniform distributions are platykurtic.

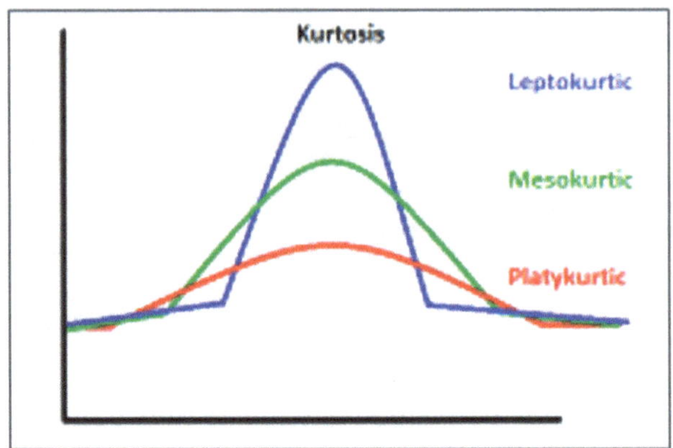

Measures of Kurtosis

The measures of kurtosis is done by moment ratio based upon second and fourth moment

$$\beta_2 = \frac{\mu_4}{\mu_2^2}$$

In a normal distribution will be equal to 3. If it is greater than 3 the curve is more peaked than normal, if less than 3 the curve is flatter at the top than normal thus

If $\beta_2 = 3$ the curve is Mesokurtic.

If $\beta_2 > 3$ the curve is Leptokurtic.

If $\beta_2 < 3$ the curve is Platykurtic.

The measure of Kurtosis is also represented by $\gamma_2. = \beta_2 - 3$

If γ_2 is positive the curve is Leptokurtic and if γ_2 is negative the curve is Platykurtic.

Objective : From a field experiment the following data are collected

N = 100, $\sum fd_x = 75$, $\sum fd_x^2 = 1800$, $\sum fd_x^3 = 4000$, $\sum fd_x^4 = 76500$

Give comments on the shape of the curve.

Theory: The shape of the curve based on the measures of skewness and kurtosis, so

$$\beta_1 = \frac{\mu_3^2}{\mu_2^3} \text{ and } \beta_2 = \frac{\mu_4}{\mu_2^2}$$

Process

Step I: First we calculate moments about origin

$$\mu_1' = \frac{\sum fd_x}{N}$$

$$= \frac{75}{100}$$

$$= 0.75$$

$$\mu_2' = \frac{\sum fd_x^2}{N}$$

$$= \frac{1800}{100}$$

$$= 18$$

$$= 40$$

$$\mu_3' = \frac{\sum fd_x^3}{N}$$

$$= \frac{76500}{100}$$

$$= 765$$

Step II: Now Calculate central moments

$$\mu_1 = 0$$

$$\mu_2 = \mu_2' - (\mu_1')^2$$

$$= 18 - (0.75)^2$$

$$= 18 - 0.5625$$

$$= 17.44$$

$$\mu_3 = \mu_3' - 3\mu_2'\mu_1' + 2(\mu_1')^3$$

$$= 0.3438$$

$$= 40 - 3*18*0.75 + 2*(0.75)^3$$

$$= 704.80$$

$$\beta_1 = \frac{\mu_3^2}{\mu_2^3}$$

$$= \frac{(0.3438)^2}{(17.44)^3}$$

$= 0.002$

$= 2.32$

Flow Diagram:

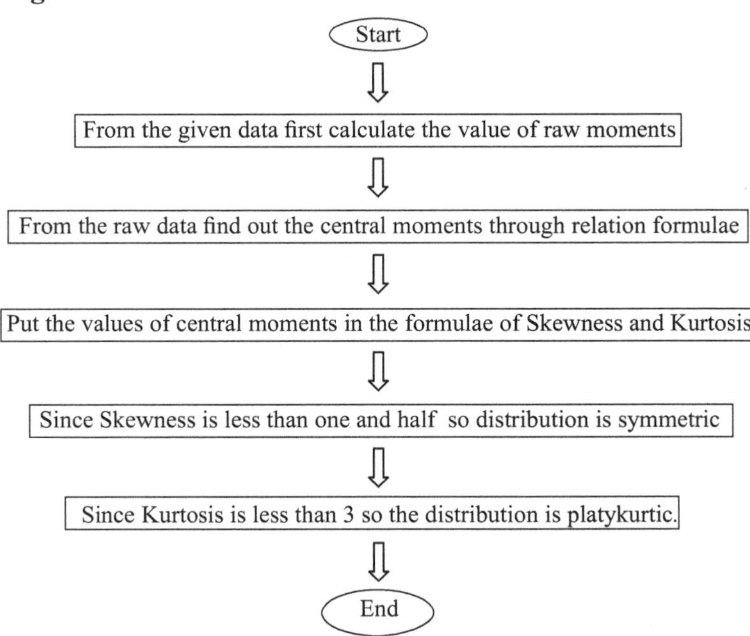

Result: From β_1 and β_2 we conclude that the distribution have symmetrical and platykurtic property.

FOR GROUPED DATA

Objective: Calculate the coefficient of kutosis through step deviation method for the following data

Class:	0-10	10-20	20-30	30-40	40-50
Frequency:	1	3	5	7	4

Theory: The coefficient of kutosis is

$$\beta_2 = \frac{\mu_4}{\mu_2^2}$$

Where μ_2 and μ_4 are the secnd and fourth moment about origin.

We will obtain μ_2 and μ_4 through step deviation method.

$$\mu_1' = \frac{\sum fd_s}{N} \; where \; d_s = \frac{x-a}{h}$$

$$\mu_2' = \frac{\sum fd_s^2}{N}$$

$$\mu_3' = \frac{\sum fd_s^3}{N}$$

$$\mu_4' = \frac{\sum fd_s^4}{N}$$

$$\mu_1 = 0$$

$$\mu_2 = \mu_2' - (\mu_1')^2$$

$$\mu_3 = \mu_3' - 3\mu_2'\mu_1' + 2(\mu_1')^3$$

$$\mu_4 = \mu_4' - 4\mu_2'\mu_1' + 6\mu_2'(\mu_1')^2 - 3(\mu_1')^4$$

Process: Step I: Here a=25 and h=10

Class	Frequency F	Mid value X	$d_s = \frac{x-25}{10}$	fd_s	fd_s^2	fd_s^3	fd_s^4
0-10	1	5	-2	-2	4	-8	16
10-20	3	15	-1	-3	3	-3	3
20-30	5	25	0	0	0	0	0
30-40	7	35	1	7	7	7	7
40-50	4	45	2	8	16	32	64
Total	**20**			**10**	**30**	**28**	**90**

Step II:

$$\mu_1 = \frac{\sum fd_s}{N}$$

$$= \frac{10}{20}$$

$$= 0.5$$

$$\mu_2 = \frac{\sum fd_s^2}{N} = \frac{30}{20} = 1.5$$

$$\mu_3 = \frac{\sum \text{fd}_s^3}{\text{N}} = \frac{28}{20} = 1.4$$

$$\mu_4 = \frac{\sum \text{fd}_s^4}{\text{N}} = \frac{90}{20} = 4.5$$

$$\mu_4 = 0$$

hence central moments are

$$\mu_2 = \left(\mu_2^{'} - (\mu_1^{'})^2\right) * h^2$$

$$= \left(1.5 - (0.5)^2\right) * 100 = 1.5$$

$$\mu_3 = (\mu_3^{'} - 3\mu_2\mu_1^{'} + 2(\mu_1^{'}))^3 * h^3$$

$$= (1.4 - (3*1.5*0.5) + 2(0.5))^3 * 10^3 = -600$$

$$\mu_4 = \left(\mu_4^{'} - 4\mu_2\mu_1^{'} + 6\mu_2^{'}(\mu_1^{'})^2 - 3(\mu_1^{'})^4\right) * h^4$$

$$= \left(4.5 - (4*1.5*0.5) + 6*1.5(0.5)^2 - 3(0.5)^4\right) * 10^4 = 37625$$

Step III: The coefficient of kutosis is

$$\beta_2 = \frac{\mu_4}{\mu_2^2}$$

$$= \frac{37625}{(125)^2}$$

$$= 2.408$$

Since $\beta_2 < 3$ the distribution is platykurtic.

Flow Diagram:

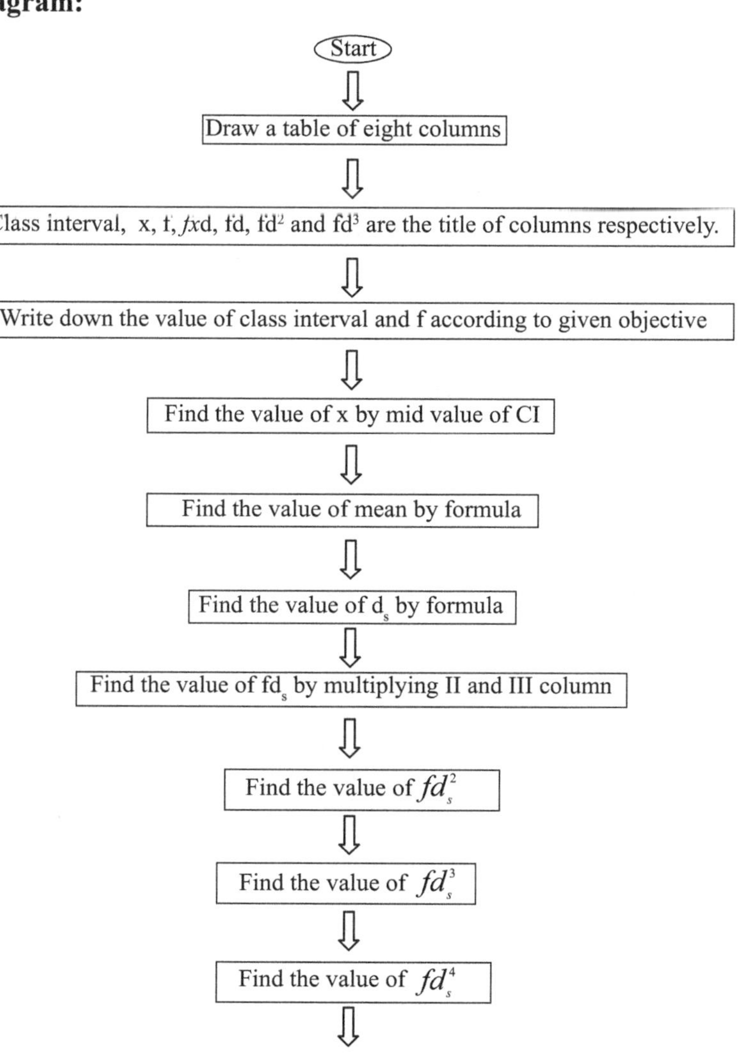

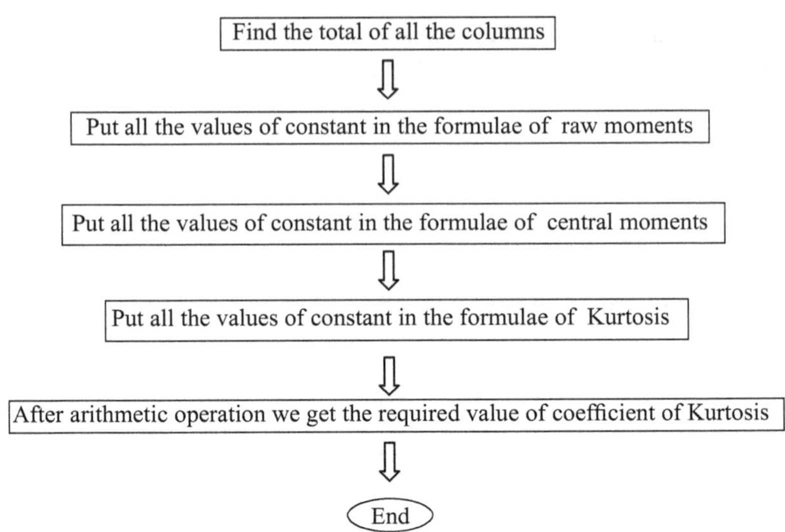

Result: From the given data we found that the distribution is platykurtic.

Related Questions

Q.1 What is Kurtosis.

Q.2 How many measures of Kurtosis?

Q.3 What is coefficient of Kurtosis?

Q.4 How many types of Kurtosis?

Q.5 What is γ_2?

17

Study of Correlation Coefficient

CORRELATION

In case of bivariate or multivariate normal distributions, we may be interested in discovering and measuring the magnitude and direction of the relationship between two or more variables. For this purpose we use correlation. The systematic interrelationship between the variables is termed as correlation. When only two variables are involved the correlation is known as simple correlation. If more than two variables are involved the correlation is known as multiple correlation. An increase in one variable may cause an increase in the other variable, or a decrease in one variable may cause a decrease in the other variable, then they are said to be positively correlated. If a decrease in one variable causes an increase in the other variable or vise versa, the variables are said to be negatively correlated.

CORRELATION COEFFICIENT

In correlation problems, first we have to investigate whether there is any relation between the variables. For this purpose we use scatter diagram but the scatter diagram will give only a vague idea about the presence or absence of correlation and the nature of correlation. It will not indicate about the strength or degree of relationship between two variables. The index of the degree of relationship between two continuous variables is known as correlation coefficient. Correlation coefficient is denoted by r and known as Pearson's Correlation coefficient.

The correlation coefficient, r is given as the ratio of covariance of the variables X and Y . Mathematically,

$$r = \frac{\frac{1}{n-1}\Sigma(x-\overline{x})(y-\overline{y})}{\sqrt{\frac{1}{n-1}\Sigma(x-\overline{x})^2}\sqrt{\frac{1}{n-1}\Sigma(y-\overline{y})^2}}$$

$$= \frac{\Sigma(x-\overline{x})y-\overline{y})}{\sqrt{\Sigma(x-\overline{x})^2}\sqrt{\Sigma(y-\overline{y})^2}}$$

This can be simplified as

$$r = \frac{\Sigma xy - \frac{\Sigma x \Sigma y}{n}}{\sqrt{\Sigma x^2 - \frac{(\Sigma x)^2}{n}}\sqrt{\Sigma y^2 - \frac{(\Sigma y)^2}{n}}}$$

The limit of correlation coefficient is lies between -1 to +1.

Assumption for correlation coefficient: The correlation coefficient is use under certain assumption

1. The variables under study are continuous random variables and they are normally distributed.
2. The relationship between the variables is linear.
3. Each pair of observation is unconnected with other pairs.

Objective: Compute Karl Pearson's coefficient of correlation between agricultural production and industrial production from the following data of index numbers of two variables:

Index no. of agricultural production	98	102	114	117	117	124	115	132	127	135
Index no. of Industrial production	112	113	117	129	139	151	153	157	175	194

Theory: The formula for Karl Pearson's Coefficient of correlation is

$$r = \frac{\Sigma xy - \frac{\Sigma x \Sigma y}{n}}{\sqrt{\Sigma x^2 - \frac{(\Sigma x)^2}{n}}\sqrt{\Sigma y^2 - \frac{(\Sigma y)^2}{n}}}$$

Where n is the no. of samples.

Process: We use the following step for find out the correlation coefficient

Step I: Draw a table of 5 columns.

Step II: X(Index no. of agricultural production) , Y(Index no. of industrial production) , XY, X^2 and Y^2 are the headings of these columns.

Step III: Find the sum of column I and II.

Step IV: Multiplying the column I and column II to get XY and find the sum of the column.

Step V: Squaring Column I to get X^2 then find the sum of the column.

Step VI: Squaring Column II to get Y^2 then find the sum of the column.

Step VII: Now put all the value of column totals in the formula of correlation coefficient to get the correlation coefficient.

X	Y	XY	X^2	Y^2
98	112	10976	9604	12544
102	113	11526	10404	12769
114	117	13338	12996	13689
117	129	15093	13689	16641
117	139	16263	13689	19321
124	151	18724	15376	22801
115	153	17595	13225	23409
132	157	20724	17424	24649
127	175	22225	16129	30625
135	194	26190	18225	37636
1181	**1440**	**172654**	**140761**	**214084**

Put all the values of sums in formula, we get

$$r = \frac{172654 - \frac{(1181*1440)}{10}}{\sqrt{140761 - \frac{(1181)^2}{10}}\sqrt{214084 - \frac{(1440)^2}{10}}}$$

$= 0.8811$

Flow Diagram:

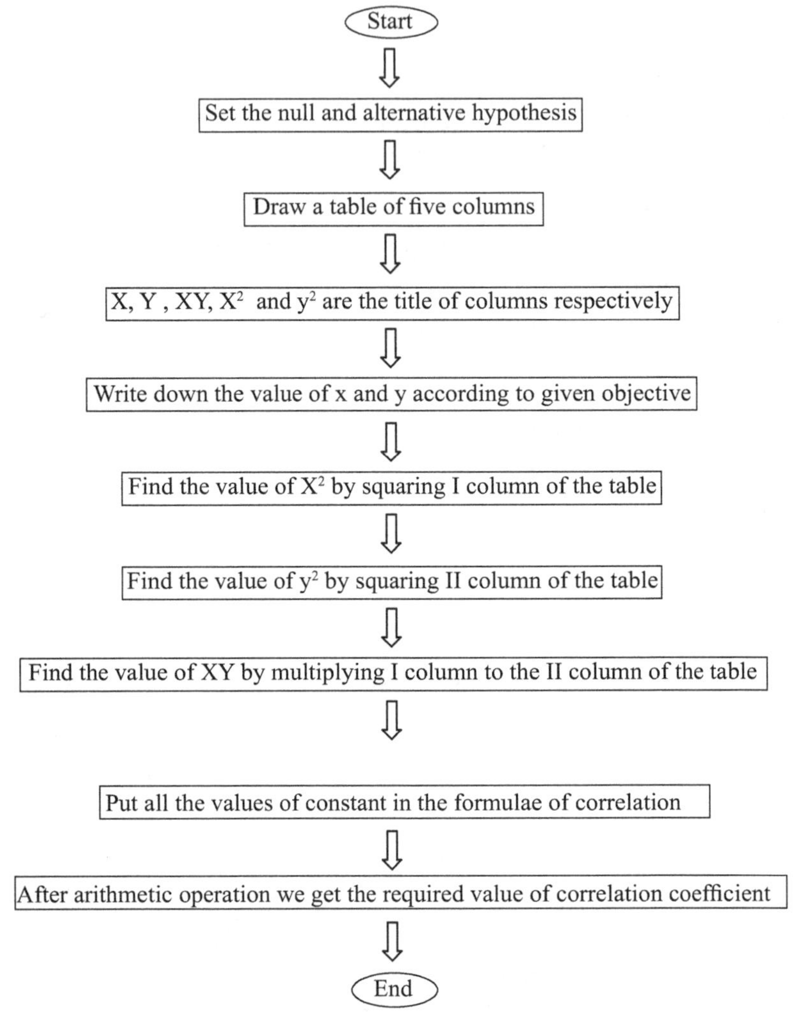

Result: The Karl Pearson's coefficient of correlation between agricultural production and industrial production is 0.8811.

Related Questions:

Q.1 What is correlation?

Q.2 What is correlation coefficient?

Q.3 What is the limit of correlation coefficient?

Q.4 Who discover correlation coefficient?

Q.5 What is scatter diagram?

18

Study of Significance of Correlation Coefficient

TEST OF SIGNIFICANCE FOR CORRELATION COEFFICIENT

Coefficient of correlation is a numeric characteristic of a population and is estimated from a sample of the bi-variate population. If ρ be the coefficient of correlation in the population and r be its sample estimate based on n pairs of observations, then standard error of r is given by

$$r = \frac{1-\rho^2}{\sqrt{n-1}}$$

If $H_0 = 0$ the value of t is calculated by the following expression

$$t = \frac{r\sqrt{n-2}}{\sqrt{1-r^2}}$$

with d.f. n-2.

If t is significant at the desired level of significance, r will be declared as significant and hypothesis is rejected.

Objective: Find out the correlation coefficient between seed yield per plant and plan height of sesamum using data in the following table apply test of significance.

Seed yield per plant	25	30	32	35	37	40	42	45
Plan Height	8	10	15	17	20	22	24	25

Theory : The formula for calculating correlation coefficient is

$$r = \frac{\frac{1}{n-1}\Sigma(x-\bar{x})(y-\bar{y})}{\sqrt{\frac{1}{n-1}\Sigma(x-\bar{x})}\sqrt{\frac{1}{n-1}\Sigma(y-\bar{y})}}$$

and test of significance is

If $H_0 = 0$, the value of t is calculated by the following expression

$$t = \frac{r\sqrt{n-2}}{\sqrt{1-r^2}}$$

With d.f. n-2.

Process: We use the following step for find out the correlation coefficient

Step I: Draw a table of 7 columns.

Step II: x (Seed yield per plant), y (Plant height), $(x-\bar{x})(y-\bar{y}),(x-\bar{x}),(y-\bar{y})$ and $(x-\bar{x})(y-\bar{y})$ are the headings of these columns.

Step III: Find the sum of column I and II.

Step IV: Find the mean of X and Y.

Step V: Calculate, $(x-\bar{x})$ and squaring them to calculate $(x-\bar{x})^2$.

Step VI: Calculate $(y-\bar{y})$ and squaring them to calculate $(y-\bar{y})^2$

Step VII: Calculate $(x-\bar{x})(y-\bar{y})$

Step VIII: Now put all the value of column totals in the formula of correlation coefficient to get the correlation coefficient.

X	Y	$X-\bar{X}$	$y-\bar{y}$	$(X-\bar{X})^2$	$(Y-\bar{Y})^2$	$(X-\bar{X})(Y-\bar{Y})$
25	8	-10.75	-9.625	115.5625	92.64063	103.4688
30	10	-5.75	-7.625	33.0625	58.14063	43.84375
32	15	-3.75	-2.625	14.0625	6.890625	9.84375
35	17	-0.75	-0.625	0.5625	0.390625	0.46875
37	20	1.25	2.375	1.5625	5.640625	2.96875
40	22	4.25	4.375	18.0625	19.14063	18.59375
42	24	6.25	6.375	39.0625	40.64063	39.84375
45	25	9.25	7.375	85.5625	54.39063	68.21875
286	**141**			**307.5**	**277.875**	**287.25**

$$\bar{X} = \frac{\sum_{i=1}^{n} X_i}{n} = \frac{286}{8} = 35.75$$

$$\bar{Y} = \frac{\sum_{i=1}^{n} y_i}{n} = 17.62$$

by the formulae

$$r = \frac{287.25/7}{\sqrt{307.5/7}\sqrt{277.88/7}}$$

$$= \frac{41.0357}{\sqrt{43.9286}\sqrt{39.6971}}$$

$$= \frac{41.0357}{(6.6279 * 6.3005)}$$

$= 0.9826$

Step IX: Test of Significance

$$t = \frac{r\sqrt{n-2}}{\sqrt{1-r^2}}$$

$$= \frac{(0.9826)\sqrt{8-2}}{\sqrt{1-(0.9826)^2}}$$

$= 12.9587$

Step X: The tabulated value of t for the 6 d.f. at 1% level of significance is 5.959 and at 5% level of significance is 2.447.

since $t_{cal} < t_{tab}$ (12.9587>2.447) then we reject the null hypothesis.

Flow Diagram:

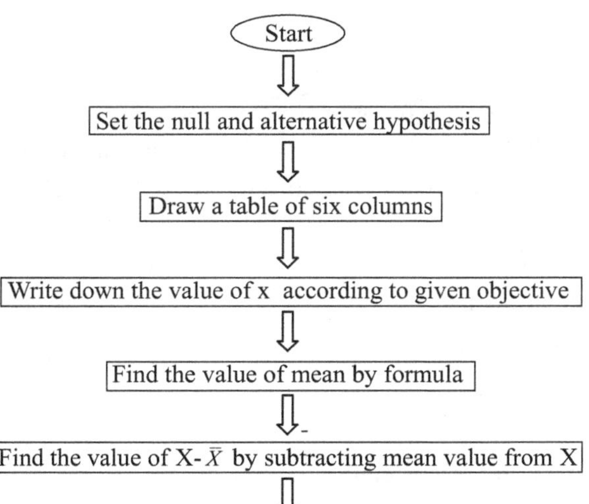

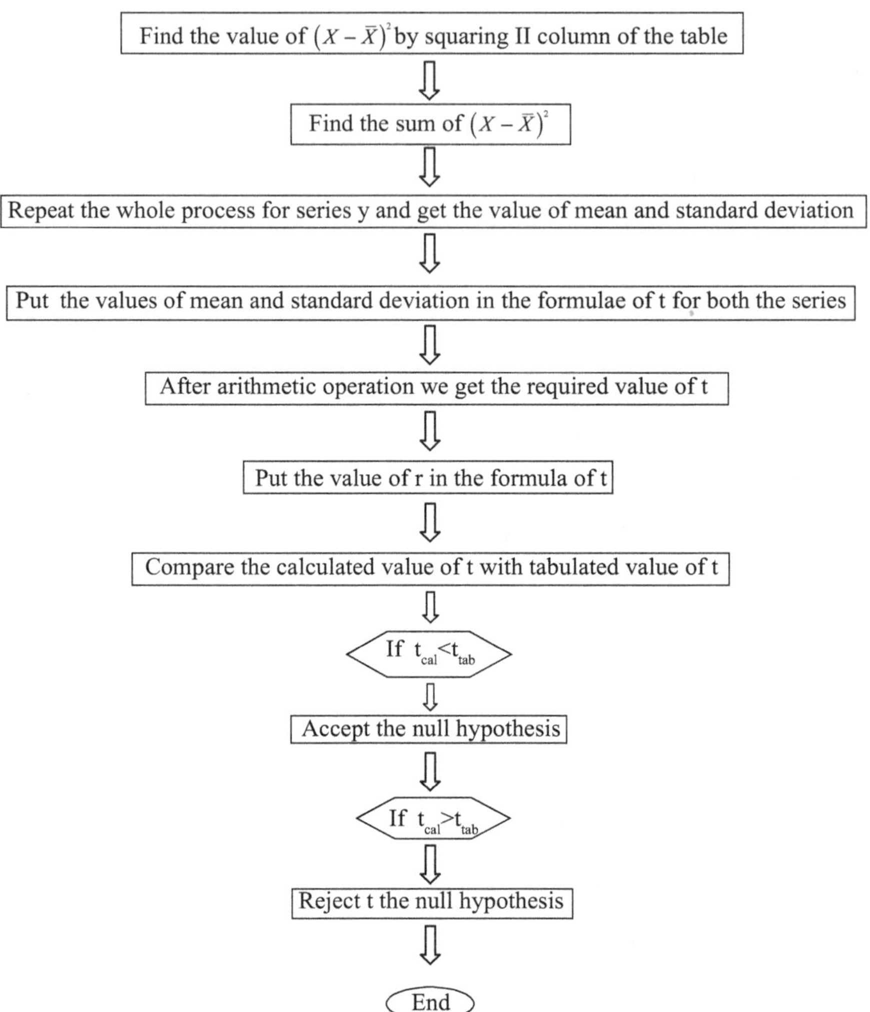

Result: The correlation coefficient between seed yield per plant and plant height of sesamum is 0.9826 and observed t is highly significant. Hence we conclude that seed yield per plant are associated with the plant height of sesamum.

Related Question

Q.1 What is significance?

Q.2 What is consistency?

Q.3 What is efficiency?

Q.4 What is unbiased?

Q.5 What is meaning of correlation significance?

19

Study of Regression Coefficient

REGRESSION

Regression means to revert or return back. The term was first introduced by Sir Francis Galton in1877. Regression technique is applicable in all those fields where two or more relative variables have the tendency to go back to the mean. Regression analysis refers to the methods by which estimates are made of the values of a variable from the knowledge of the values of one or more other variables and to the measurement of the errors involved in this estimation process.

In statistical modeling, regression analysis is a statistical process for estimating the relationships among variables. It includes many techniques for modeling and analyzing several variables, when the focus is on the relationship between a dependent variable and one or more independent variables. Regression analysis helps one understand how the typical value of the dependent variable (or 'criterion variable') changes when any one of the independent variables is varied, while the other independent variables are held fixed. Most commonly, regression analysis estimates the conditional expectation of the dependent variable given the independent variables – that is, the average value of the dependent variable when the independent variables are fixed.

Regression Lines

Regression line is the device used for estimating the value of one variable from the value of the other consists of a line through the points in such a manner as to represents the average relationship between the two variables

A linear regression line has an equation of the form $Y = a + bX$, where X is the independent variable and Y is the dependent variable. The slope of the line is b, and a is the intercept (the value of y when $x = 0$), which is also called Y on X. The other one is X on y of the form $X = a + bY$, where Y is the independent variable and X is the dependent variable. The slope of the line is b, and a is the intercept (the value of y when $x = 0$)

Regression Equations

The regression equations express the regression lines. As there are two regression lines, so there are two regression equations.

The regression equation X on Y is

$$X - \bar{X} = b_{XY}\left(Y - \bar{Y}\right)$$

where $b_{XY} = r\frac{\sigma_X}{\sigma_Y}$ regression coefficient X on Y

The regression equation Y on X is

$$Y - \bar{Y} = b_{YX}\left(X - \bar{X}\right)$$

where $b_{YX} = r\frac{\sigma_Y}{\sigma_X}$ regression coefficient Y on X.

Objective: In an experiment with eight fields planted to corn. Four fields having no nitrogen fertilizer and four having 80 pounds of nitrogen fertilizer. The resulting corn yields are shown in the table in bushels per acre:

Field	Nitrogen (Pound)	Corn yield (bushels per acre)
1.	0	12
2.	0	36
3.	0	6
4.	0	18
5.	80	128
6.	80	112
7.	80	112
8.	80	76

(a) Compute a linear regression equation Y on X .

(b) Predict corn yield for a field treated with 60 pounds of fertilizer.

Theory: The regression equation Y on X is

$$Y - \bar{Y} = b_{YX}\left(X - \bar{X}\right)$$

where $b_{YX} = r\frac{\sigma_Y}{\sigma_X}$ regression coefficient Y on X.

$$= \frac{\Sigma d_x d_Y - \frac{\Sigma d_x \Sigma d_Y}{N}}{\Sigma d_X^2 - \frac{(d_X)^2}{N}}$$

Process: Step I:

Field	Nitrogen X	$d_x = X-\bar{X}$	d_X^2	Corn Yield Y	$d_y = Y-\bar{Y}$	d_Y^2	$d_x d_y$
1	0	-40	1600	12	-48	2304	1920
2	0	-40	1600	36	-24	960	960
3	0	-40	1600	6	-54	2160	2160
4	0	-40	1600	18	-42	1680	1680
5	80	40	1600	128	68	4624	2720
6	80	40	1600	112	52	2704	2080
7	80	40	1600	112	52	2704	2080
8	80	40	1600	76	16	256	2080
Total	**320**		**12800**	**500**		**17848**	**14240**

$$\bar{X} = \frac{\sum X}{n} = \frac{320}{8} = 40$$

$$\bar{Y} = \frac{\sum Y}{n} = \frac{500}{8} = 62.5$$

Step II:

$$b_{YX} = \frac{\sum d_x d_Y - \frac{\sum d_x \sum d_Y}{n}}{\sum d_X^2 - \frac{(d_X)^2}{n}}$$

$$= \frac{14240}{12800}$$

$$= 1.1125$$

Step III: Regression equation of corn yield Y on fertilizer X

$$Y - \bar{Y} = b_{YX}\left(X - \bar{X}\right)$$

$$Y - 62.5 = 1.1125(X - 40)$$

$$Y - 62.5 = 1.1125X - 44.5$$

$$Y = 1.1125X + 18$$

Step IV: The value of Y when X= 60

= 1.1125*60+18

– 84.75 bushels per acre

Flow Diagram

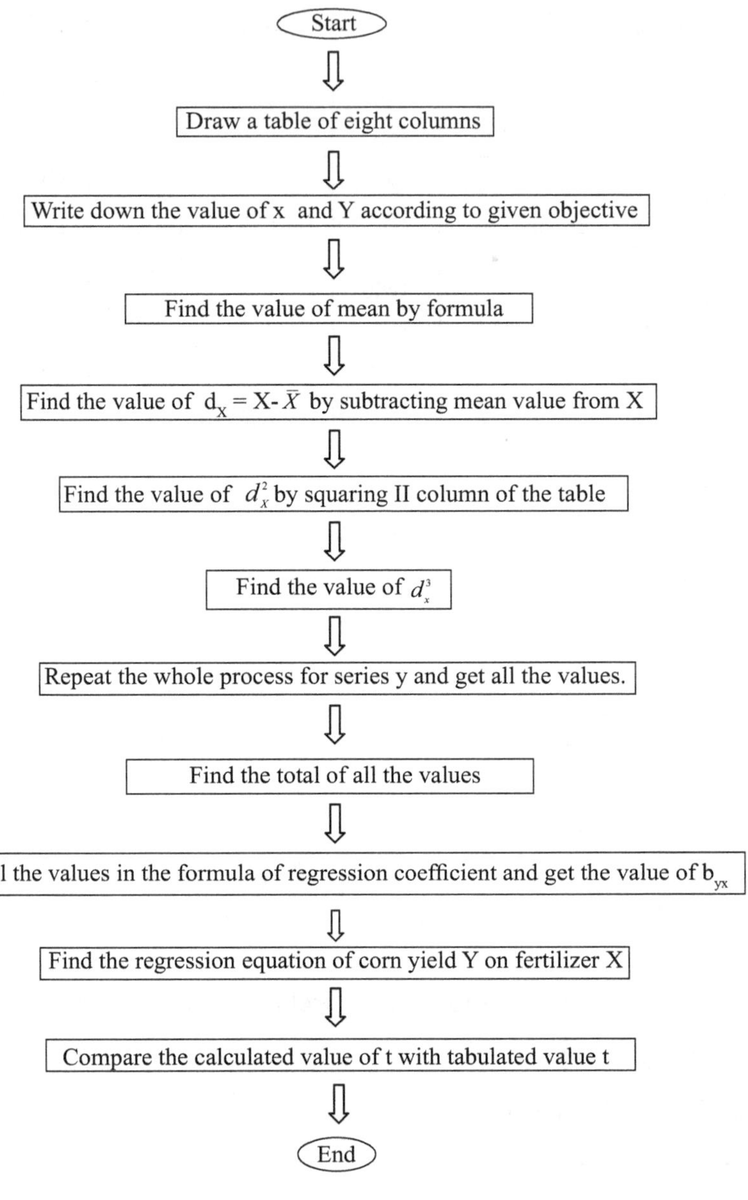

Result: From the given data we found that

(i) Regression equation of corn yield Y on fertilizer is Y= 1.1125 x t 18

(ii) Corn yield for a field treated with 60 pounds of fertilizer = 84.75 bushels per acre.

Related Questions

Q.1 What is regression?

Q.2 What is regression coefficient?

Q.3 What is difference between regression and correlation?

Q.4 What is regression equation?

Q.5 What is line of best fit?

20

Study of t-test for Single Mean

t TEST FOR SINGLE MEAN

A ***t*-test** is any statistical hypothesis test in which the test statistic follows a Student's *t* distribution if the null hypothesis is supported. It can be used to determine if two sets of data are significantly different from each other, and is most commonly applied when the test statistic would follow a normal distribution if the value of a scaling term in the test statistic were known. When the scaling term is unknown and is replaced by an estimate based on the data, the test statistic (under certain conditions) follows a Student's *t* distribution.

$$t = \frac{\overline{x} - \mu}{s / \sqrt{n}}$$

Where

μ = Population Mean

$\overline{x}$ = Sample Mean

s = Sample standard deviation

n = No. of sample observation

if $t_{cal} > t_{tab}$ then the difference is significant and null hypothesis rejected at 5% or 1% level of significance.

if $t_{tab} < t_{cal}$ then the difference is non- significant and null hypothesis accepted at 5% or 1% level of significance.

Objective: Six boys are selected at random from a school and their marks in Mathematics are found to be 63, 63, 64, 66, 60, 68 out of 100. In the light of these marks, discuss the general observation that the mean marks in Mathematics in the school were 66.

Theory: t-test for sample mean is

$$t = \frac{\overline{x} - \mu}{s / \sqrt{n}}$$

Process: Step I:

H_0 =Mean marks in Mathematics were 66 .

Vs H_1= Mean marks in Mathematics were not 66.

Step II: Given that μ= 66 , and n=6 then

Marks	d =x- $\overline{x}$	d^2
63	-1	1
63	-1	1
64	0	0
66	2	4
60	-4	16
68	4	16
384	**0**	**38**

Sample mean $\bar{X} = \frac{384}{6}$

$=64$

Standard deviation of sample

$$S = \sqrt{\frac{\sum d^2}{n-1}}$$

$$= \sqrt{\frac{38}{5}}$$

$$= 2.756$$

Step III:

$$t = \frac{\bar{x} - \mu}{s/\sqrt{n}}$$

$$= \frac{64 - 66}{2.756/\sqrt{6}}$$

t = 1.7777

and d.f. = 6-1= 5

t_{tab} = 2.571 for 5 d.f. and 5% level of significance.

Since $t_{cal} < t_{tab}$, we accept null hypothesis.

Flow Diagram

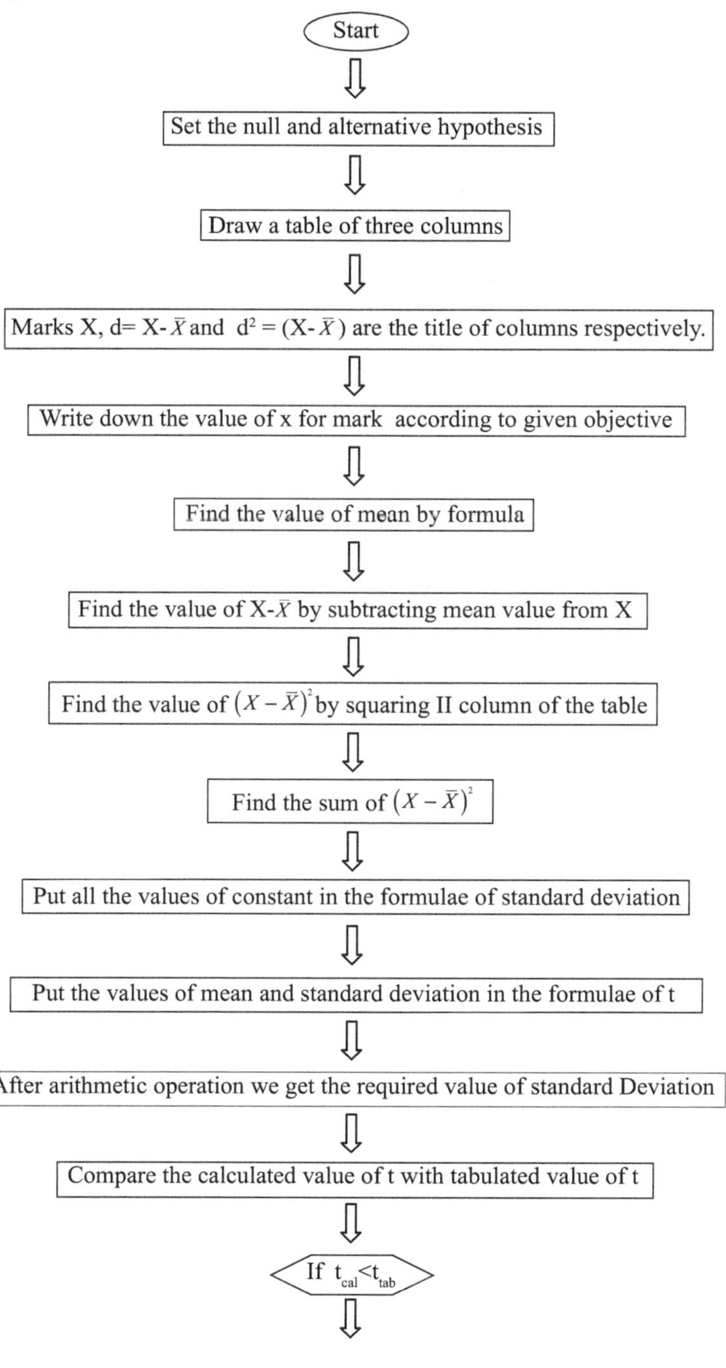

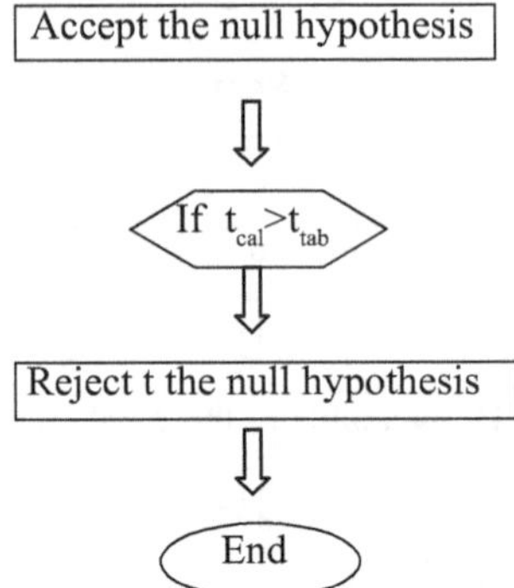

Result: Here we accept null hypothesis this means that the mean marks in Mathematics would be 66.

Related Questions

(1) When t test is applicable?

(2) What is assumption of t test?

(3) Who discovered t test?

(4) What is hypothesis?

(5) How many types of hypothesis?

21

Study of t-Test for Two Sample Mean

t TEST FOR TWO SAMPLE MEAN:

Comparison of two sample means $\overline{x}$ and $\overline{y}$ assumed to have been obtained on the basis of random samples of sizes n_1 and n_2 from the same population which is assumed to be normal.

The approximate test is given by (under H_0: = against H_1: ≠)

$$t = \frac{\overline{x} - \overline{y}}{s\sqrt{\frac{1}{n_1} + \frac{1}{n_2}}}$$

where

$$\overline{X} = \frac{\sum X_i}{n} \text{ and } \overline{Y} = \frac{\sum Y_i}{n}$$

$$s^2 = \frac{1}{n_1 + n_2 - 2}\left[\sum_{i=1}^{n_1}(x_i - \overline{x})^2 + \sum_{i=1}^{n_2}(y_i - \overline{y})^2\right]$$

$$= \frac{n_1 s_1^2 + n_2 s_2^2}{n_1 + n_2 - 2}$$

follows Student's t statistics with $n_1 + n_2 - 2$d.f.

Objective : Two kinds of manure applied to 15 plots of one acres; other condition remaining the same. The yields (in quintals) are given below:

Manure I:	14	20	34	48	32	42	30	44
Manure II:	31	18	22	28	40	26	45	

Examine the significance of the difference between the mean yields due to the application of different kinds of manure.

Theory: Here we use t test for difference of mean

$$t = \frac{\bar{x} - \bar{y}}{s\sqrt{\frac{1}{n_1} + \frac{1}{n_2}}}$$

$\bar{X}$ and $\bar{Y}$ are the sample mean of I and II sample.

Process: Step I:

H_0 : There is no significance difference between two the mean yields due to the application of different kinds of manure.

V_s H_1 : There is significance difference between two the mean yields due to the application of different kinds of manure.

Step II:

Manure I	$(x-\bar{x})$	$(x-\bar{x})^2$	Manure II	$(y-\bar{y})$	$(y-\bar{y})^2$
14	-19	361	31	+1	1
20	-13	169	18	-12	144
34	+1	1	22	-8	64
48	+15	225	28	-2	4
32	-1	1	40	+10	100
42	+9	81	26	-4	16
30	-3	9	45	+15	225
44	+11	121			
264		**968**	**210**		**554**

Step III:

$$\bar{X} = \frac{\Sigma X_i}{n} \quad \bar{Y} = \frac{\Sigma Y_i}{n}$$

$$= \frac{264}{8} \quad = \frac{210}{7}$$

$$= 33 \quad = 30$$

Step IV:

$$s^2 = \frac{1}{n_1 + n_2 - 2}\left[\sum_{i=1}^{n_1}(x_i - \bar{x})^2 + \sum_{i=1}^{n_2}(y_i - \bar{y})^2\right]$$

$$\frac{1}{8+7-2}968+554$$

$= 117.07$

$s = 10.82$

Step V:

$$t = \frac{\bar{x}-\bar{y}}{s\sqrt{\frac{1}{n_1}+\frac{1}{n_2}}}$$

$$\frac{33-30}{10.82\sqrt{\frac{1}{8}+\frac{1}{7}}}$$

$= 0.54$

$\text{d.f.} = n_1 + n_2 - 2 = 13$

the tabulated value of t for 13 d. f at 5% d. f. is 2.16. since $t_{cal} < t_{tab}$ then we accept null hypothesis.

Flow Diagram:

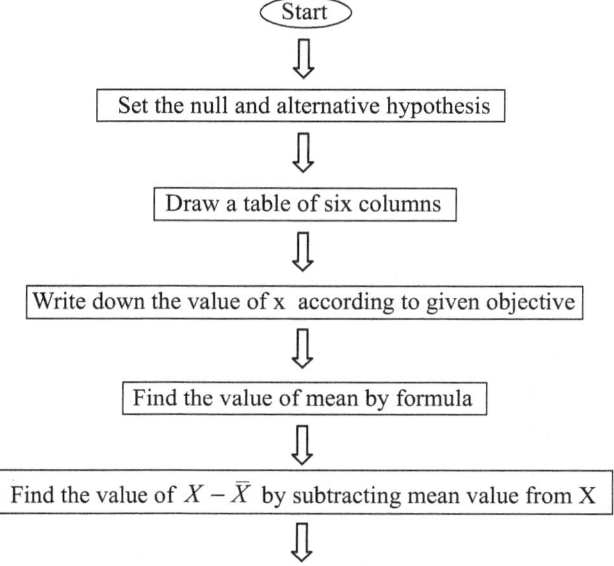

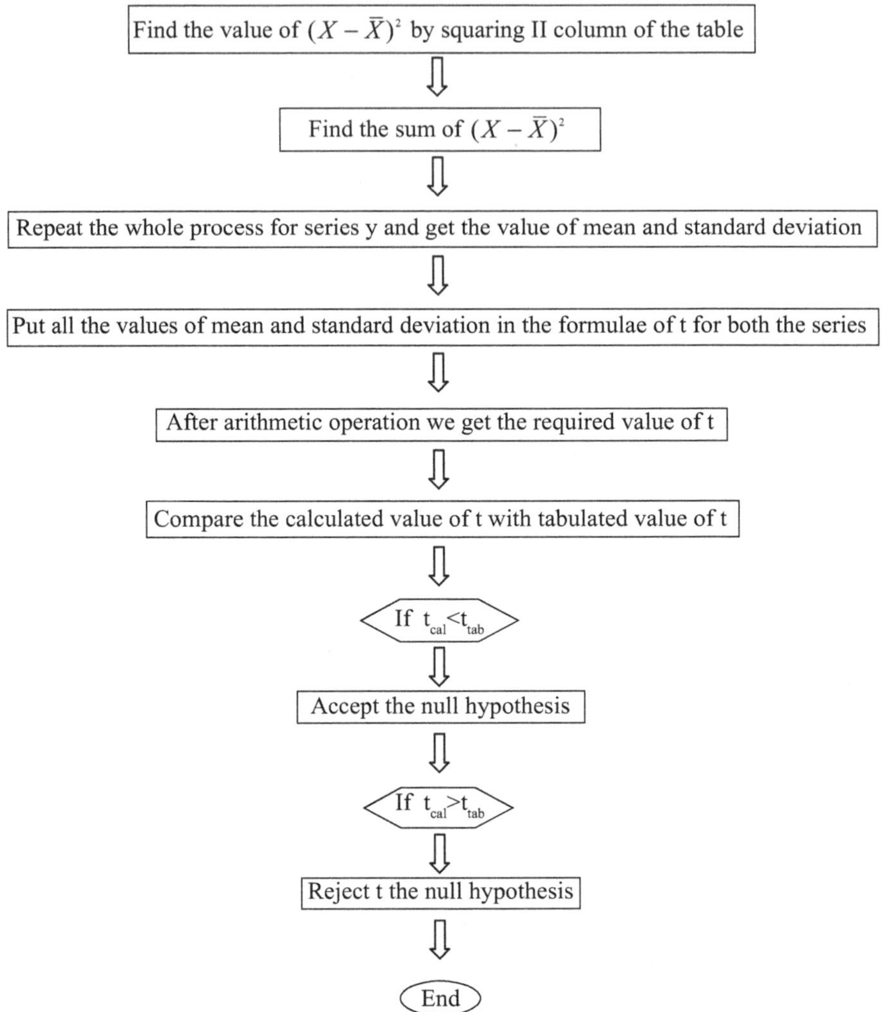

Result: Since $t_{cal} < t_{tab}$ then we conclude that there is no significance difference between two mean yields due to the application of different kinds of manure.

Related Question

1. When difference of mean test is useful?
2. What is degree of freedom?
3. When the hypothesis accepted?
4. When the hypothesis rejected?
5. What is the assumption of difference of mean test?

22

Study of t -Test for Paired Observation

t TEST FOR PAIRED OBSERVATION

This test is used for testing whether two series of paired observations are generated from the same population on the basis of the difference in their sample means. The approximate test is given by

$$t = \frac{\bar{d}}{s/\sqrt{n}}$$

where

$$\bar{d} = \frac{\sum_{i=1}^{n} d_i}{n} \text{ and } s^2 = \frac{\sum_{i=1}^{n}\left(d_i - \bar{d}\right)^2}{n-1}$$

$d_i = x_i - y_i$ being the difference of the ith observation in the two sample follows student's t-distribution with n-1 d.f.

Objective: A certain stimulus administrated to each 12 patients resulted in the following change in blood pressure

5 2 8 -1 3 0 -2 1 5 0 4 6

can it be concluded that the stimulus will in general be accompanied by an increase in blood pressure?

Theory: Student's t-distribution with n-1 d.f. is

$$t = \frac{\bar{d}}{s/\sqrt{n}}$$

where

$$\bar{d} = \frac{\sum_{i=1}^{n} d_i}{n} \text{ and } s^2 = \frac{\sum_{i=1}^{n}\left(d_i - \bar{d}\right)^2}{n-1}$$

$d_i = x_i - y_i$

Process: Step I:

Hypothesis

H_0: There is no increase in blood pressure by stimulus.

Vs H_1: There is an increase in blood pressure by stimulus.

Step II:

Change in Blood Pressure (d_i)	$d_i - \bar{d}$	$(d - \bar{d})^2$
5	2.4167	5.840439
2	-0.5833	0.340239
8	5.4167	29.34064
-1	-3.5833	12.84004
3	0.4167	0.173639
0	-2.5833	6.673439
-2	-4.5833	21.00664
1	-1.5833	2.506839
5	2.4167	5.840439
0	-2.5833	6.673439
4	1.4167	2.007039
6	3.4167	11.67384
31		**104.9167**

Step III:

$$\bar{d} = \frac{\sum_{n}^{i=1} d_i}{n} \text{ and } s^2 = \frac{\sum_{i=1}^{n}\left(d_i - \bar{d}\right)^2}{n-1}$$

$$= \frac{31}{12} \qquad = \frac{104.9167}{12-1}$$

$$= 2.583 \qquad = 9.5379$$

$$= 2 \qquad s = 3.0883$$

Step IV:

$$t = \frac{2.583}{3.0883 / \sqrt{12}}$$

$$= 2.8976$$

degree of freedom = 12-1 = 11

Step V

t_{tab} at 5% level of significance and 11 degree of freedom is 2.201.

since $t_{cal} > t_{tab}$, hence null hypothesis is rejected.

Flow Diagram

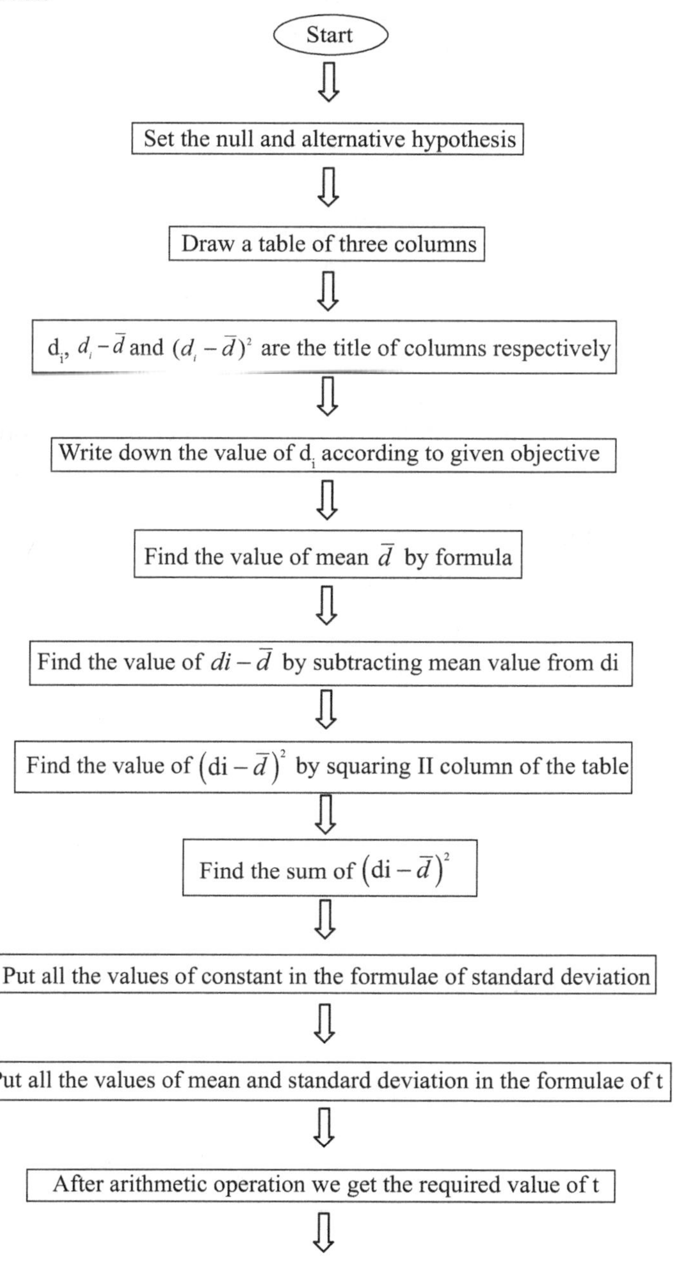

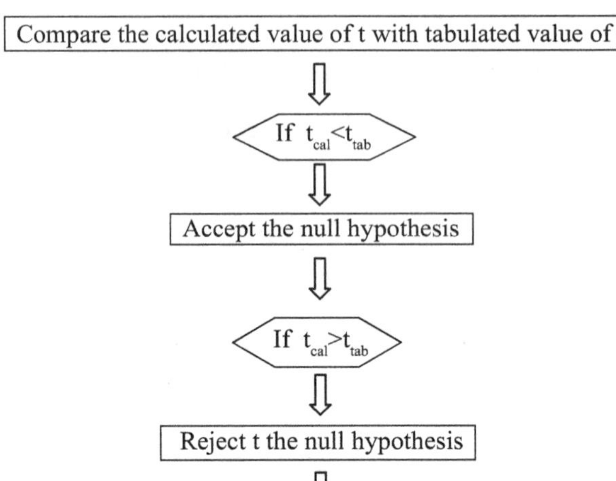

Result: since $t_{cal} > t_{tab}$, hence null hypothesis is rejected so we concluded that stimulus will in general be accompanied by an increase in blood pressure.

Related Questions

Q.1 When paired test is applicable?

Q.2 In paired t test what is d?

Q.3 Assumption of paired t test?

Q.4 What is α ?

Q.5 What is β?

23

Study of F Test

F TEST (VARIANCE RATIO TEST)

F distribution is applied in several tests of significance relating to the equality of two sampling variances drawn on the basis of independent samples from a normal population. The approximate test is

$$\text{Variance Ratio (F)} = \frac{\text{Larger estimate of variance}}{\text{Smaller estimate of variance}}$$

$$= \frac{s_1^2}{S_2^2}$$

where $s_1^2 = \frac{\sum_{i=1}^{n_1}(x_i - \bar{x})^2}{n_1 - 1}$ and $S_2^2 = \frac{\sum_{i=1}^{n_2}(x_2 - \bar{x})^2}{n_2 - 1}$

Follows F distribution with n_1-1 and n_2-1 d.f..

Objective: For a random sample of 10 pigs, fed on diet A, the increase in weight in a certain period were:

10 6 16 17 13 12 8 14 15 9 lbs.

for an other random sample of12 pigs, fed on diet B, the increase in the same were:

7 13 22 15 12 14 18 8 21 23 10 17 lbs.

Show that the estimate of the population variance from the two samples are not significantly different. (For n_1 = 11, n_2= 9 the 5% value of F is 3.112)

Theory: F distribution with n_1-1 and n_2-1 d.f. then

$$\text{Variance Ratio (F)} = \frac{\text{Larger estimate of variance}}{\text{Smaller estimate of variance}}$$

$$= \frac{s_1^2}{S_2^2}$$

where $s_1^2 = \frac{\sum_{i=1}^{n_1}(x_i - \bar{x})^2}{n_1 - 1}$ and $S_2^2 = \frac{\sum_{j=1}^{n2}(Y_j - \bar{Y})^2}{n_2 - 1}$

Process: Step I:

Hypothesis:

H_0: There is no significance difference between the estimates of the population variances of two samples.

$V_s H_1$: There is significance difference between the estimates of the population variances of two samples.

Step II:

Sample I			Sample II		
x_i	$(x_i-\bar{x})$	$(x_i-\bar{x})^2$	y_i	$(y_i-\bar{y})$	$(y_i-\bar{y})^2$
10	-2	4	7	-8	64
6	-6	36	13	-2	4
16	4	16	22	7	49
17	5	25	15	0	0
13	1	1	12	-3	9
12	0	0	14	-1	1
8	-4	16	18	3	9
14	2	4	8	-7	49
15	3	9	21	6	36
9	-3	9	23	8	64
			10	-5	25
			17	2	4
120		**120**	**180**		**314**

Step III:

$$\bar{x} = \frac{\Sigma X_i}{n} \quad \bar{y} = \frac{\Sigma Y_i}{n}$$

$$= \frac{120}{10} \qquad = \frac{180}{12}$$

$$= 12 \qquad = 15$$

Step IV:

$$s_1^2 = \frac{\Sigma_{i=1}^{n_1} (x_i - \bar{x})^2}{n_1 - 1} \text{ and } S_2^2 = \frac{\Sigma_{j=1}^{n_2} (y;-\bar{y})^2}{n_2 - 1}$$

$$= \frac{20}{10-1} \qquad = \frac{180}{180-1}$$

$$= 13.33 \qquad = 28.55$$

Step V:

$$F = \frac{S_1^2}{S_2^2}$$

$$= \frac{28.55}{13.33}$$

$$= 2.1418$$

Since $F_{cal} < F_{tab}$ then null hypothesis is accepted.

Flow Diagram:

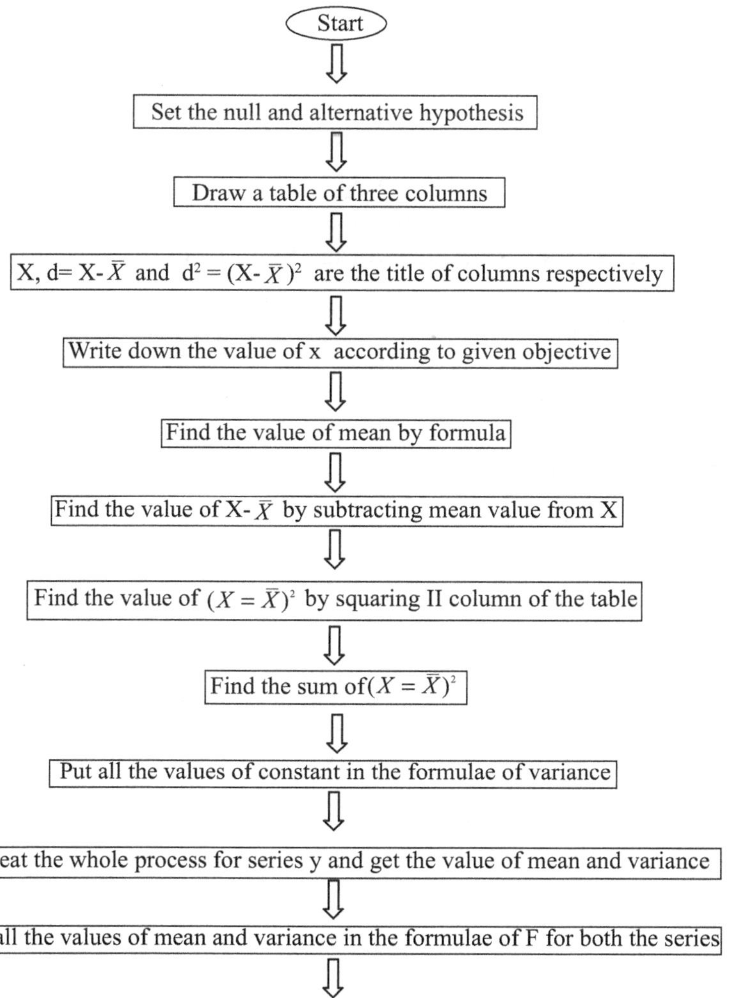

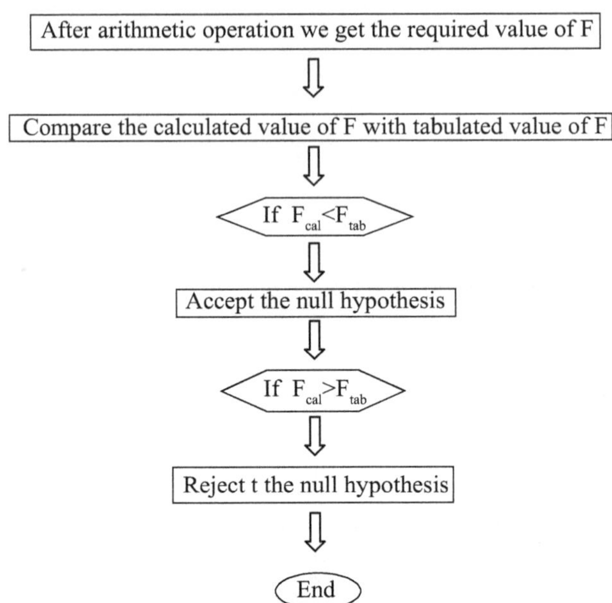

Result: Since $F_{cal} < F_{tab}$ the null hypothesis is accepted and we conclude that the population variance from the two samples are not significantly different.

Related Question

Q.1 What is variance test?

Q.2 Assumption of F Test?

Q.3 When F test is applicable?

Q.4 Who discovered F test?

Q.5 Is F test applicable for testing of mean?

24

Study of Standard Normal Test

STANDARD NORMAL DISTRIBUTION

If μ = 0 and σ = 1, the distribution is called the standard normal distribution or the unit normal distribution, and a random variable with that distribution is a standard normal deviate. The approximate test is given by

$$Z = \frac{x - \mu}{\sigma}$$

Where

x = Expected Mean

μ = Population Mean

σ = Population Variance

Objective: As a result of tests on 20,000 electric bulbs manufactured by a company it was found that the life time of the bulbs was normally distributed with an average of 2040 hours and a standard of 60 hours. On the basis of this information estimate the number of bulbs that are expected to burn for (a) more than 2150 hours (b) less than 1960 hours. The are under normal curve is given

Z:	1.23	1.33	1.43	1.53	1.63	1.73	1.83
A:	0.3907	0.4082	0.4236	0.4484	0.4482	0.4664	0.4464

Theory: We use Z test

$$Z = \frac{x - \mu}{\sigma}$$

Where x = Expected Mean

μ = Population Mean

σ = Population Variance

Process: Step I: here μ = 2040 hours, σ = 1960 hours and N = 20,000 then

Step II:

(a) for more than 2150 hours

$$Z = \frac{110}{1.83}$$

= 1.83

Area right to 1.83 is = 0.5-0.4664

= 0.0336

Therefore no. of bulbs that are expected to burn = .0336*20000 = 672.

Step III:

(b) less than 1960 hours

$$Z = \frac{1960 - 2040}{= -1.33}$$

= -1.33

Area left to -1.33 is = 0.5-0.4882

= 0.0918

Therefore no. of bulbs that are expected to burn = .0918*20000 = 1836.

Related Questions

Q.1 What is large test?

Q.2 When large test is useful?

Q.3 What is Z?

Q.4 Why this test is called standard normal?

Q.5 When any distribution is called Normal?

25

Study of χ^2 Test

CHI- SQUARE TEST (χ^2 TEST)

The chi square test is used to determine if the two attributes are independent of each other. Chi square is a measure to evaluate the difference between observed frequencies and expected frequencies to examine whether the difference so obtained is due to a chance factor or due to sampling error.

In a contingency table if each attribute is divided into two classes it is known as 2×2 contingency table. When one attribute is divided into two classes and another into r or c resultant contingency table is known as r×2 or 2×c contingency table. Here, r denotes the number of rows and c the number of columns. For such data, the statistical hypothesis under test is that the two attribute are independent of one another. To test this hypothesis we use the test statistic

$$\chi^2 = \sum \frac{(O_i - E_i)^2}{E_i}$$

Where,

O_i = Observed Frequency

E_i = Expected Frequency

Objective: In order to determine the possible effect of a chemical treatment on the rate of germination of cotton seeds a pot culture experiment was conducted. Five hundred chemically treated seeds and 1500 untreated seeds were sown. The results are given in below table

	Germinated	Not Germinated
Chemically Treated	31	469
Untreated	185	1315

Does the chemical treatment improve the germination rate of cotton seeds? Test your result at 5% level of significance.

Theory: Here we use χ^2 test

$$\chi^2 = \sum \frac{(O_i - E_i)^2}{E_i}$$

Where,

O_i = Observed Frequency

E_i = Expected Frequency

Process: Step I:

Hypothesis

H_0: The chemical treatment does not improve the germination rate of cotton seeds i.e. $O_i = E_i$

V_s

H_1: The chemical improves the germination rate of cotton seeds i.e. $O_i = E_i$.

Step II: Observed Frequencies:

	Germinated	Not Germinated	Total
Chemically Treated	31	469	500
Untreated	185	1315	1500
Total	216	1784	2000

Step III : Expected Frequencies

For germinated

$$= \frac{500 \times 216}{2000} = 54 \qquad = \frac{1500 \times 216}{2000} = 162$$

For not germinated

$$= \frac{500 \times 1784}{2000} = 446 \qquad = \frac{1500 \times 1784}{2000} 1338$$

Expected Frequencies:

	Germinated	Not Germinated	Total
Chemically Treated	54	162	500
Untreated	446	1338	1500
Total	**216**	**1784**	**2000**

Step IV:

Observed Frequencies (O_i)	Expected Frequencies(E_i)	Difference (O_i-E_i)	Square of differences $(O_i-E_i)^2$	$(O_i-E_i)^2/E_i$
31	54	-23	529	9.796
469	446	23	529	1.186
185	162	23	529	3.265
1315	1338	-23	529	0.395
Total				**14.642**

Step V:

Degree of Freedom= (2-1)(2-1) = 1

Table values for 1 degree of freedom at 5% level of significance = 3.841

Since

$\chi^2_{cal} = 14.642$ and $\chi^2_{tab} = 3.841$

$\chi^2_{cal} > \chi^2_{tab}$, we reject the null hypothesis.

Flow Diagram:

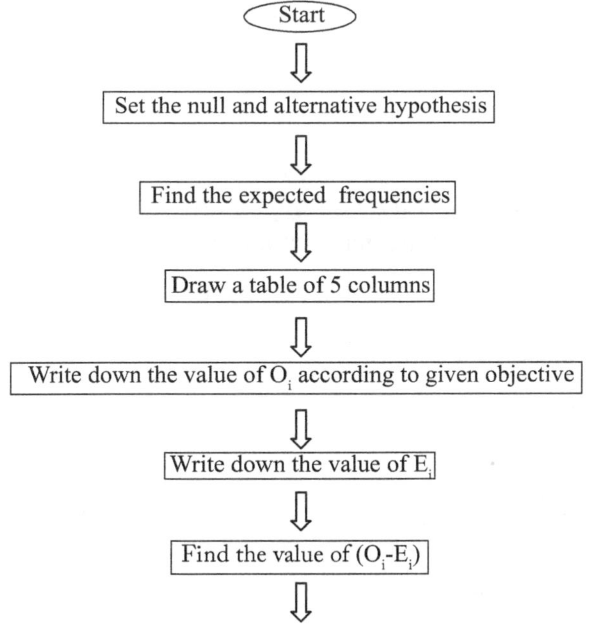

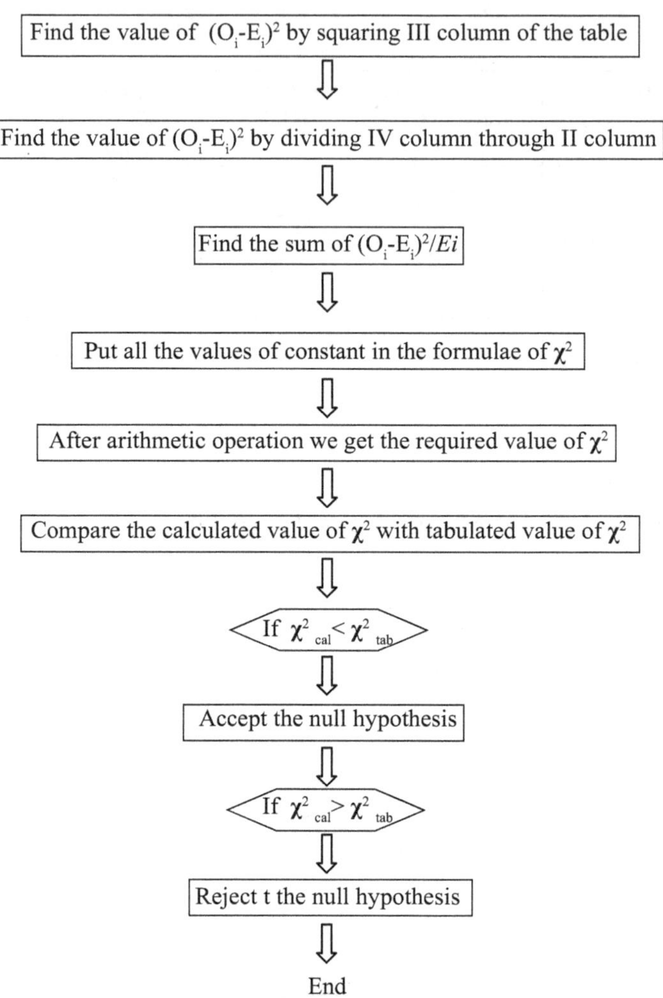

Result: $\chi^2_{cal} > \chi^2_{tab}$ we reject the null hypothesis. It may be conclude that the chemical treatment improve the germination rate of cotton seeds.

Related Questions

Q.1 What is non parametric test?

Q.2 What is parametric test?

Q.3 Define χ^2

Q.4 Assumption for χ^2 test.

Q.5 When χ^2 test is applicable?

26

Study of Yates Correction for χ^2 Test

The chi square is a continuous distribution based on the assumption that the ultimate class frequency is not less than 5 otherwise the value of χ^2 will be over-estimated and may cause too many rejections of null hypothesis. Therefore, to maintain the continuity of χ^2 and to draw the correct inference in 2×2 we apply a correction which is known as Yates's correction for continuity.

The Yates correction consists of the following steps:

1. Add 0.5 of the cell frequency which is least
2. Adjust the remaining cell frequencies in such a way that the row and column totals are not changed.

It can be shown that this correction will result in the formula

$$\chi^2 = \frac{\left[|ad - bc| - \frac{n}{2}\right]^2 n}{c_1 c_2 r_1 r_2}$$

Objective: An experiment on the effect of immunization of goats against a disease was conducted. Two batches, of 25 animals, were taken. One batch was inoculated and the other was not inoculated. Then both the batches were exposed to the infection of the disease. The frequencies of dead and survived animals were observed in both the batches. The results are given in below table. From the results test whether the inoculation is effective against the disease?

	Dead	Survived
Inoculated	16	9
Not inoculated	5	20

Test your result at 1% level of significance.

Theory: The formula with Yate's correction is as follows

$$\chi^2 = \frac{\left[|ad - bc| - \frac{n}{2}\right]^2 n}{c_1 c_2 r_1 r_2}$$

Where 2*2 contingency table is

$$\begin{vmatrix} a & b \\ c & d \end{vmatrix}$$

c_1 = First Column r_1 = First Row

c_2 = Second Column r_2 = Second Row

Process: Step I: Here the one cell frequency is 5, so we use Yate's correction.

Hypothesis:

H_0: Inoculation is not effective against the disease.

Vs

H_1: Inoculation is effective against the disease.

Step II:

	Dead	Survived	Total
Inoculated	16	9	25
Not inoculated	5	20	25
Total	**21**	**29**	**50**

Step III: By using Yates correction we have

$$\chi^2 = \frac{\left[|ad - bc| - \frac{n}{2}\right]^2 n}{c_1 c_2 r_1 r_2}$$

$$= \frac{\left[|16 \times 20 - 9 \times 5| - \frac{50}{2}\right]^2 50}{21 \times 29 \times 25 \times 25}$$

$= 8.210$

Step IV:

The critical value of χ^2 for 1 d.f. and $\alpha = 0.01$ is 6.635 i.e.

$\chi^2_{cal} = 8.210$ and $\chi^2_{tab} = 6.635$

$\chi^2_{cal} > \chi^2_{tab}$, we reject the null hypothesis.

Flow Diagram:

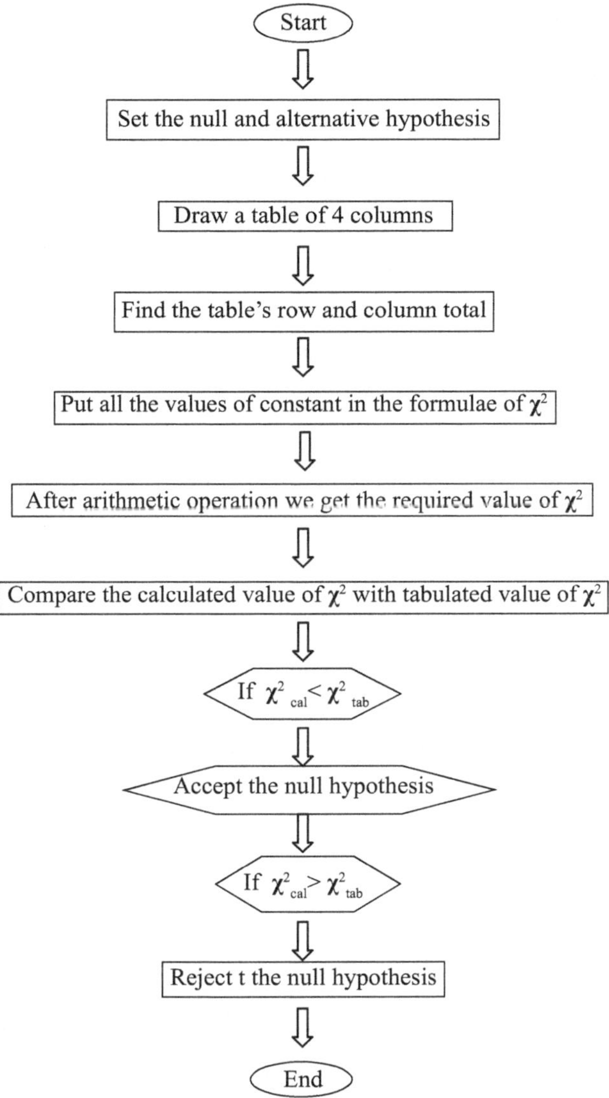

Result: $\chi^2_{cal} > \chi^2_{tab}$, we reject the null hypothesis. It may be conclude that inoculation is effective against the disease.

Related Questions

Q.1 What is contingency table?

Q.2 Why it is useful?

Q.3 When Yates correction is applicable?

Q.4 What are the conditions where chi square test is applicable?

Q.5 What is “goodness of fit”?

27

Study of One Way Analysis of Variance

ANALYSIS OF VARIANCE

Analysis of Variance (ANOVA) is a hypothesis-testing technique used to test the equality of two or more population (or treatment) means by examining the variances of samples that are taken. ANOVA allows one to determine whether the differences between the samples are simply due to random error (sampling errors) or whether there are systematic treatment effects that cause the mean in one group to differ from the mean in another.

Most of the time ANOVA is used to compare the equality of three or more means, however when the means from two samples are compared using ANOVA it is equivalent to using a t-test to compare the means of independent samples.

ANOVA is based on comparing the variance (or variation) between the data samples to variation within each particular sample. If the between variation is much larger than the within variation, the means of different samples will not be equal. If the between and within variations are approximately the same size, then there will be no significant difference between sample means.

Assumptions of ANOVA

(i) All populations involved follow a normal distribution.

(ii) All populations have the same variance (or standard deviation).

(iii) The samples are randomly selected and independent of one another.

Since ANOVA assumes the populations involved follow a normal distribution, ANOVA falls into a category of hypothesis tests known as parametric tests. If the populations involved did not follow a normal distribution, an ANOVA test could not be used to examine the equality of the sample means. Instead, one would have to use a non-parametric test (or distribution-free test), which is a more general form of hypothesis testing that does not rely on distributional assumptions.

One Way Analysis of Variance

The one-way analysis of variance (ANOVA) is used to determine whether there are any statistically significant differences between the means of two or more independent (unrelated) groups (it used when there are a minimum of three, rather than two groups). For example, we can use a one-way ANOVA to understand whether exam performance differed based on test anxiety levels amongst students, dividing students into three independent groups (e.g., low, medium and high-stressed students). Also, it is important to realize that the one-way ANOVA is an **omnibus** test statistic and cannot tell you which specific groups were statistically significantly different from each other; it only tells us that at least two groups were different.

Objective: In order to find out the yielding abilities of five verities of sesamum an experiment was conducted in the greenhouse using a CRD with four pots per variety. The results are given in the table below

Variety	Seed yield of Seasamum g/pot			
V_1	25	21	21	18
V_2	25	28	24	25
V_3	24	24	16	21
V_4	20	17	16	19
V_5	14	15	13	11

Analyze the data and state your conclusions

Theory: For one way analysis fixed effect model is

$$x_{ij} = \mu + \alpha_i + \epsilon_{ij}$$

Where

x_{ij} = j[th] unit of the ith treatment

μ is the general mean effect given by

$\mu = \frac{\sum_{i=1}^{k} n_i \mu_i}{N}$ is the fixed effect due to the ith treatment

α_i is the effect of the ith treatment given by

$$\alpha_i = \mu_i - \mu$$

$$\sum_{i=1}^{k} n_i \alpha_i = 0$$

ϵ_{ij} is the error effect due to chance

Assumptions for the model:

(i) All the observations are independent.

(ii) Different effects are additive in nature.

(iii) $\in_{ij}$ are iid N(0,)

1. Correction Factor

$$CF = \frac{(GT)^2}{n}$$

2. Total Sum of Square(TSS)

$$\sum_{i,j}^{k,ni} x_{ij}^2 - CF$$

3. Treatment sum of square

$$SSE \sum_{i=1}^{k} \sum_{j=1}^{ni} (\bar{x} - \bar{x})^2 = (\frac{\sum T_i^2}{n})CF$$

4. Sum of square due to error

$$SSE \sum_{i=1}^{k} \sum_{j=1}^{ni} (\bar{x} - \bar{x})^2 = TSS\text{-}SST$$

ANOVA Table

Source of Variation	Sum of Square	Df	Mean Sum of Square	Variance Ratio
Treatment	SST	K-1	MSST=SST/k-1	F=MSST/MSSE
Error	SSE	N-K	MSSE = SSE/N-K	
Total	TSS	N-1		

Process: We use the following step for the analysis of ANOVA

Step I: Compute correction factor through formulae

Step II: Compute total sum of square through formulae

Step III: Compute treatment sum of square through formulae

Step IV: Find sum of square due to error

Step V: Prepare ANOVA table

Step I:

Null Hypothesis:

H_0: There is no difference between varieties of yield

Vs

H_1: There is difference between varieties of yield

Variety	Seed yield of Seasamum g/pot				Total	Mean
V_1	25	21	21	18	85	21.2
V_2	25	28	24	25	102	25.5
V_3	24	24	16	21	85	21.2
V_4	20	17	16	19	72	18.0
V_5	14	15	13	11	53	13.2
Total					**397**	

$$CF = \frac{(397)^2}{20} = 7880.4$$

Step II:

Total sum of square = $(25)^2+(21)^2+\ldots(13)^2+(11)^2$ - CF

= 8307-7880.45

= 426.55

Step III:

Variety sum of square = $(85)^2+(102)^2+(85)^2+(72)^2+(53)^2$-CF

= 331.30

Step IV:

Error sum of square = 426.55 - 331.30

= 95.25

Step V:

ANOVA Table

Source of Variation	Sum of Square	df	Mean Sum of Square	Variance Ratio
Treatment	331.30	4	82.82	13.04
Error	95.25	15	6.35	
Total				

Step VI : The tabulated value of tFfor the 15 d.f. at 5% level of significance is 2.131. Since $t_{cal} > t_{tab}$, then we reject null hypothesis.

Flow Diagram:

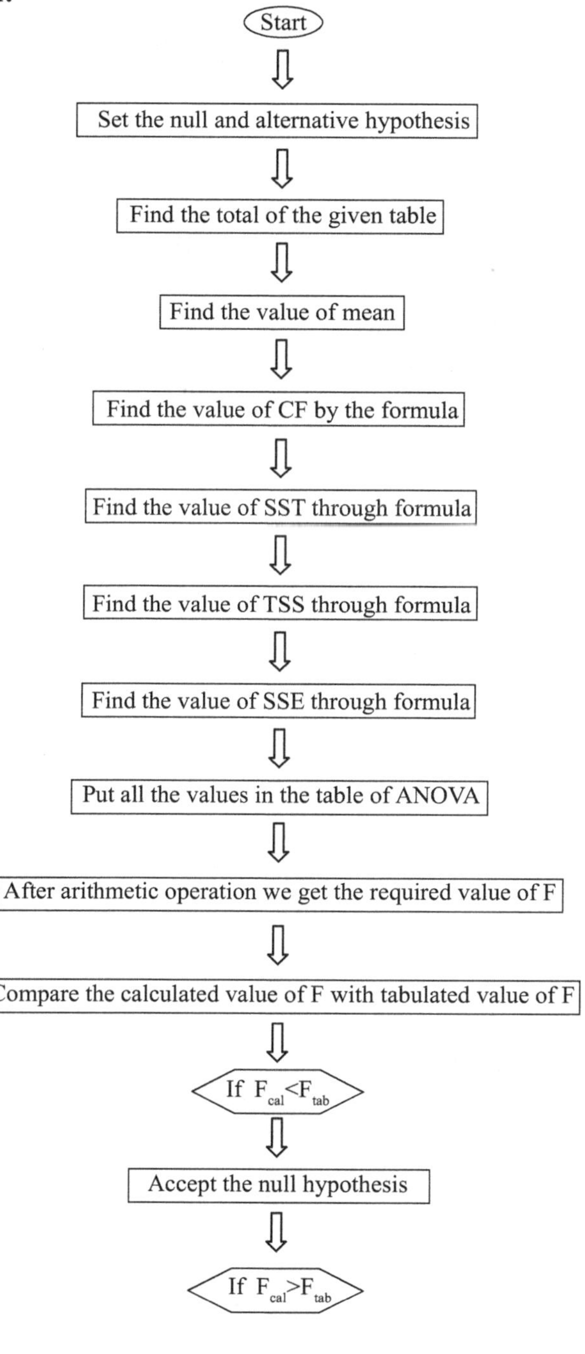

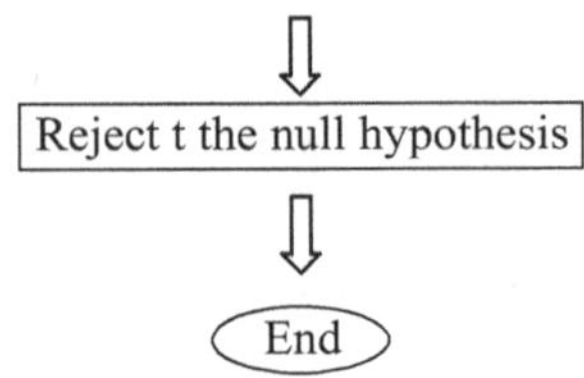

Result: Since we reject the null hypothesis so we conclude that there is significance difference between verities of mean.

Related Question

Q.1 What is design of experiment?

Q.2 What are the principles of design of experiment?

Q.3 What is ANOVA?

Q.4 What is assumption of ANOVA?

Q.5 When one way ANOVA is useful?

28

Study of Two Ways Analysis of Variance

The two-way ANOVA compares the mean differences between groups that have been split on two independent variables (called factors). The primary purpose of a two-way ANOVA is to understand if there is an interaction between the two independent variables on the dependent variable. For example, you could use a two-way ANOVA to understand whether there is an interaction between gender and educational level on test anxiety amongst university students, where gender (males/females) and education level (undergraduate/postgraduate) are your independent variables, and test anxiety is your dependent variable. Alternately, you may want to determine whether there is an interaction between physical activity level and gender on blood cholesterol concentration in children, where physical activity (low/moderate/high) and gender (male/female) are your independent variables, and cholesterol concentration is your dependent variable.

The interaction term in a two-way ANOVA informs you whether the effect of one of your independent variables on the dependent variable is the same for all values of your other independent variable (and vice versa). For example, is the effect of gender (male/female) on test anxiety influenced by educational level (undergraduate/postgraduate)? Additionally, if a statistically significant interaction is found, you need to determine whether there are any "simple main effects", and if there are, what these effects are (we discuss this later in our guide).

Assumptions

- The populations from which the samples were obtained must be normally or approximately normally distributed.
- The samples must be independent.
- The variances of the populations must be equal.
- The groups must have the same sample size.

Hypotheses

There are three sets of hypothesis with the two-way ANOVA. The null hypotheses for each of the sets are given below.

1. The population means of the first factor are equal. This is like the one-way ANOVA for the row factor.
2. The population means of the second factor are equal. This is like the one-way ANOVA for the column factor.
3. There is no interaction between the two factors. This is similar to performing a test for independence with contingency tables.

Factors

The two independent variables in a two-way ANOVA are called factors. The idea is that there are two variables, factors, which affect the dependent variable. Each factor will have two or more levels within it, and the degrees of freedom for each factor is one less than the number of levels.

Treatment Groups

Treatment Groups are formed by making all possible combinations of the two factors. For example if the first factor has 3 levels and the second factor has 2 levels, then there will be 3x2=6 different treatment groups

Main Effect

The main effect involves the independent variables one at a time. The interaction is ignored for this part. Just the rows or just the columns are used, not mixed. This is the part which is similar to the one-way analysis of variance. Each of the variances calculated to analyze the main effects are like the between variances

Interaction Effect

The interaction effect is the effect that one factor has on the other factor. The degree of freedom here is the product of the two degrees of freedom for each factor.

Within Variation

The Within variation is the sum of squares within each treatment group. You have one less than the sample size (remember all treatment groups must have the same sample size for a two-way ANOVA) for each treatment group. The total number of treatment groups is the product of the number of levels for each factor. The within variance is the within variation divided by its degrees of freedom. The within group is also called the error.

F-Tests

There is an F-test for each of the hypotheses, and the F-test is the mean square for each main effect and the interaction effect divided by the within variance. The numerator degrees of freedom come from each effect, and the denominator degrees of freedom is the degrees of freedom for the within variance in each case.

Objective: Five varieties of wheat are tested in a randomized block design with four replications. The data pertaining to yield in kg per plot are given below. Analyse the experimental yield and write your conclusions.

Varieties	Replication			
	I	II	III	IV
V_1	5	4	3	4
V_2	6	8	4	6
V_3	4	6	5	5
V_4	9	10	8	9
V_5	12	10	13	11

Given that $F_{(4,12)}$ at 0.05= 3.26

Theory:

1. Correction Factor

$$CF = \frac{(GT)^2}{n}$$

2. Total Sum of Square(TSS)

$$\sum_{i,j}^{k,ni} x_{ij}^2 - CF$$

3. Treatment sum of square

$$SST = \Sigma_{i=1}^{k} \Sigma_{j=1}^{ni} (\bar{x}_i - \bar{x}_{...})^2 = (\frac{\Sigma T_i^2}{n}) - CF$$

4. Block sum of square

$$SSB = \Sigma_{i=1}^{k} \Sigma_{j=1}^{ni} (\bar{x}_i - \bar{x}_{...})^2 = (\frac{\Sigma B_i^2}{k}) - CF$$

5. Sum of square due to error

$$SSE = \sum_{i}^{k} \sum_{j=1}^{ni} \left(\bar{x}_{ij} - \bar{x}_{i}\right)^{2} = TSS-SST-SSB$$

ANOVA Table

Source of Variation	Sum of Square	Df	Mean Sum of Square	Variance Ratio
Treatment	SST	K-1	MSST=SST/k-1	F=MSST/ MSSE
Block	SSB	n-1	MSSB=SSB/n-1	
Error	SSE	(n-1) (k-1)	MSSE = SSE/N-K	F=MSSB/ MSSE
Total	TSS	N-1		

Process: We use the following step for the analysis of ANOVA

Step I: Compute correction factor through formulae

Step II: Compute total sum of square through formulae

Step III: Compute treatment sum of square through formulae

Step IV: Compute block sum of square through formulae

Step V: Find sum of square due to error

Step VI: Prepare ANOVA table

Null Hypothesis:

H_0: There is no difference between varieties of yield

Vs

H_1: There is difference between varieties of yield

Varieties	Replication				Total	Mean
	I	II	III	IV		
V_1	5	4	3	4	16	4
V_2	6	8	4	6	24	6
V_3	4	6	5	5	20	5
V_4	9	10	8	9	36	9
V_5	12	10	13	11	46	11.5
Total	36	38	33	35	142	

$$CF = \frac{(142)^2}{20} = 1008.2$$

Total sum of square = $(5)^2+(4)^2+\ldots(13)^2+(11)^2$ - CF

= 171.8

Variety sum of square = $(16)^2+(14)^2+(20)^2+(36)^2+(46)^2$-CF

= 152.8

Block sum of square = $(36)^2+(38)^2+(33)^2+(35)^2$-CF

= 2.6

Error sum of square = 171.8 - 152.8 - 2.6

= 16.4

ANOVA Table

Source of Variation	Sum of Square	df	Mean Sum of Square	Variance Ratio
Treatment	152.8	4	38.2	27.96
Block	2.6	3	0.83	
Error	16.4	12	1.36	3.26
Total	171.8			

The tabulated value of F for the (4,12) d.f. at 5% level of significance is 3.26. since $f_{cal} > F_{tab}$, then we reject null hypothesis.

Flow Diagram:

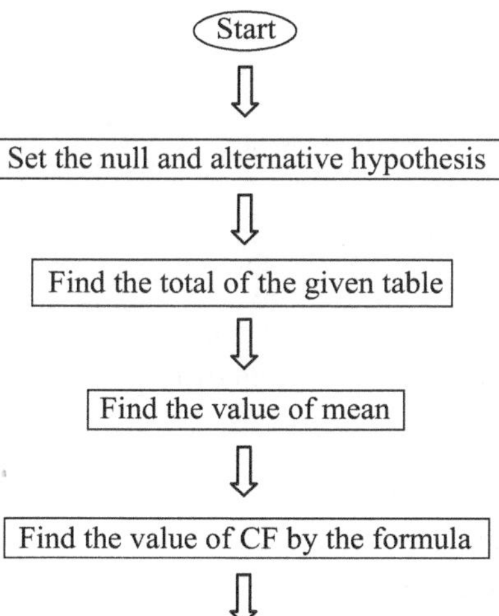

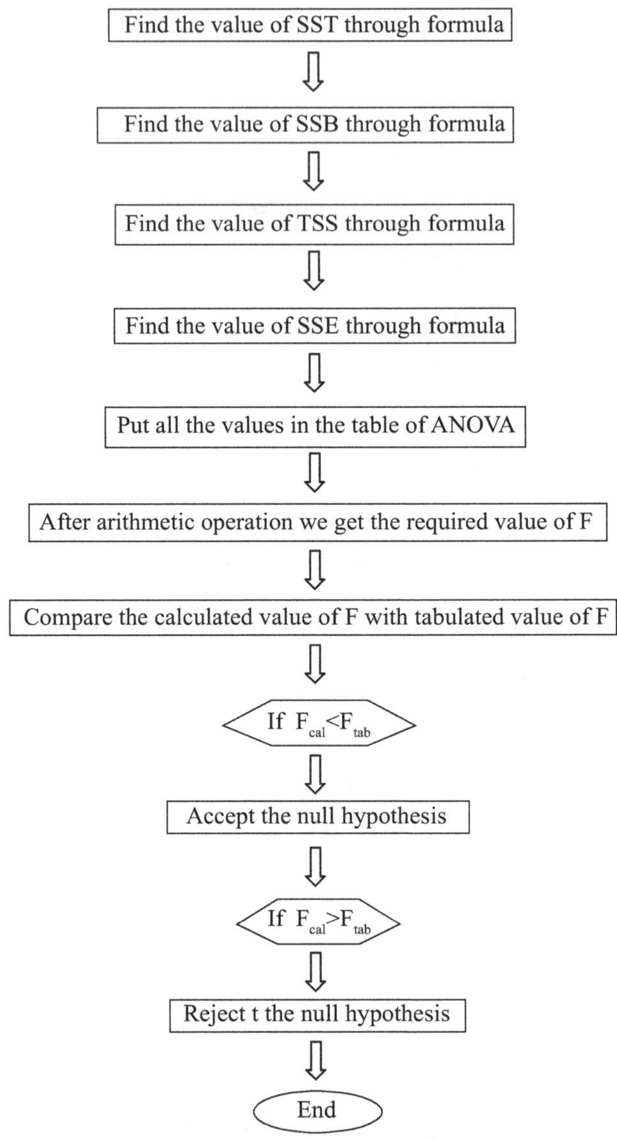

Result: Since we accept the null hypothesis so we conclude that there is significance difference between verities of mean.

Related Questions

Q.1 What is Two Way ANOVA?
Q.2 When it is Useful?
Q.3 Assumptions of two way ANOVA?
Q.4 What is main effect?
Q.5 What is interaction effect?

29

Simple Random Sampling

A simple random sample (SRS) of size n consists of n individuals from the population chosen in such a way that every set of n individuals has an equal chance to be the sample actually selected.

A simple random sample is a subset of a statistical population in which each member of the subset has an equal probability of being chosen. An example of a simple random sample would be the names of 25 employees being chosen out of a hat from a company of 250 employees. In this case, the population is all 250 employees, and the sample is random because each employee has an equal chance of being chosen.

Selection of a Simple Random Sample

Random sample refers to that method of sample selection in which every item has an equal chance of being selected. But the random sample does not depend upon the method of selection only but also on the size and nature of the population. Random sample can be obtained by any method:

(i) Lottery system method

(ii) Random Number method

Objective: draw a random sample (without replacement) of size 30 from a population of size 10000.

Theory: A sample of Tippet's table of numbers is

2952	3392	7979	3170
4167	1545	7203	3100
2370	3408	3563	6913
0560	1112	6008	4433
2754	1405	7002	8816
6641	9792	5911	5624
9524	1396	5356	2993
7483	2672	1089	7691
5246	6107	8126	8796
9143	9025	6111	9446

Procedure: Step I: We identify each of 10000 units by a serial number.

Step II: We will pick up the first 30 digits from Tippet's random number table which have a value up to 10000.

2952	3392	7979	3170
4167	1545	7203	3100*
2370	3408	3563	6913
0560	1112	6008*	4433
2754	1405	7002	8816
6641	9792*	5911	5624
9524	1396	5356	2993
7483	2672	1089	-
5246	6107	8126	-
9143	9025	6111*	-

*Not selected because last digit already selected.

Result: The required random numbers are

52	92	03	16
67	45	63	24
70	08	02	93
60	12	11	
54	05	56	
41	96	89	
24	72	26	
83	07	70	
46	25	13	
43	79	33	

Related questions

Q.1 What is sampling?

Q.2 What is random sampling?

Q.3 What is random Number?

Q.4 How many types of sampling?

Q.5 What is population and sample?

Appendix

Values of the t-distribution (two-tailed)

Df	A	0.80	0.90	0.95	0.98	0.99	0.995	0.998	0.999
	P	0.20	0.10	0.05	0.02	0.01	0.005	0.002	0.001
1		3.078	6.314	12.706	31.820	63.657	127.321	318.309	636.619
2		1.886	2.920	4.303	6.965	9.925	14.089	22.327	31.599
3		1.638	2.353	3.182	4.541	5.841	7.453	10.215	12.924
4		1.533	2.132	2.776	3.747	4.604	5.598	7.173	8.610
5		1.476	2.015	2.571	3.365	4.032	4.773	5.893	6.869
6		1.440	1.943	2.447	3.143	3.707	4.317	5.208	5.959
7		1.415	1.895	2.365	2.998	3.499	4.029	4.785	5.408
8		1.397	1.860	2.306	2.897	3.355	3.833	4.501	5.041
9		1.383	1.833	2.262	2.821	3.250	3.690	4.297	4.781
10		1.372	1.812	2.228	2.764	3.169	3.581	4.144	4.587
11		1.363	1.796	2.201	2.718	3.106	3.497	4.025	4.437
12		1.356	1.782	2.179	2.681	3.055	3.428	3.930	4.318
13		1.350	1.771	2.160	2.650	3.012	3.372	3.852	4.221
14		1.345	1.761	2.145	2.625	2.977	3.326	3.787	4.140
15		1.341	1.753	2.131	2.602	2.947	3.286	3.733	4.073
16		1.337	1.746	2.120	2.584	2.921	3.252	3.686	4.015
17		1.333	1.740	2.110	2.567	2.898	3.222	3.646	3.965
18		1.330	1.734	2.101	2.552	2.878	3.197	3.610	3.922
19		1.328	1.729	2.093	2.539	2.861	3.174	3.579	3.883
20		1.325	1.725	2.086	2.528	2.845	3.153	3.552	3.850
21		1.323	1.721	2.080	2.518	2.831	3.135	3.527	3.819
22		1.321	1.717	2.074	2.508	2.819	3.119	3.505	3.792
23		1.319	1.714	2.069	2.500	2.807	3.104	3.485	3.768
24		1.318	1.711	2.064	2.492	2.797	3.090	3.467	3.745
25		1.316	1.708	2.060	2.485	2.787	3.078	3.450	3.725
26		1.315	1.706	2.056	2.479	2.779	3.067	3.435	3.707
27		1.314	1.703	2.052	2.473	2.771	3.057	3.421	3.690
28		1.313	1.701	2.048	2.467	2.763	3.047	3.408	3.674
29		1.311	1.699	2.045	2.462	2.756	3.038	3.396	3.659
30		1.310	1.697	2.042	2.457	2.750	3.030	3.385	3.646
31		1.309	1.695	2.040	2.453	2.744	3.022	3.375	3.633
32		1.309	1.694	2.037	2.449	2.738	3.015	3.365	3.622
33		1.308	1.692	2.035	2.445	2.733	3.008	3.356	3.611
34		1.307	1.691	2.032	2.441	2.728	3.002	3.348	3.601
35		1.306	1.690	2.030	2.438	2.724	2.996	3.340	3.591
36		1.306	1.688	2.028	2.434	2.719	2.991	3.333	3.582
37		1.305	1.687	2.026	2.431	2.715	2.985	3.326	3.574
38		1.304	1.686	2.024	2.429	2.712	2.980	3.319	3.566
39		1.304	1.685	2.023	2.426	2.708	2.976	3.313	3.558

40		1.303	1.684	2.021	2.423	2.704	2.971	3.307	3.551
42		1.302	1.682	2.018	2.418	2.698	2.963	3.296	3.538
44		1.301	1.680	2.015	2.414	2.692	2.956	3.286	3.526
46		1.300	1.679	2.013	2.410	2.687	2.949	3.277	3.515
48		1.299	1.677	2.011	2.407	2.682	2.943	3.269	3.505
50		1.299	1.676	2.009	2.403	2.678	2.937	3.261	3.496
60		1.296	1.671	2.000	2.390	2.660	2.915	3.232	3.460
70		1.294	1.667	1.994	2.381	2.648	2.899	3.211	3.435
80		1.292	1.664	1.990	2.374	2.639	2.887	3.195	3.416
90		1.291	1.662	1.987	2.369	2.632	2.878	3.183	3.402
100		1.290	1.660	1.984	2.364	2.626	2.871	3.174	3.391
120		1.289	1.658	1.980	2.358	2.617	2.860	3.160	3.373
150		1.287	1.655	1.976	2.351	2.609	2.849	3.145	3.357
200		1.286	1.652	1.972	2.345	2.601	2.839	3.131	3.340
300		1.284	1.650	1.968	2.339	2.592	2.828	3.118	3.323
500		1.283	1.648	1.965	2.334	2.586	2.820	3.107	3.310
		1.282	1.645	1.960	2.326	2.576	2.807	3.090	3.291

Standard Normal Table

z	.00	.01	.02	.03	.04	.05	.06	.07	.08	.09
.0	.5000	.5040	.5080	.5120	.5160	.5199	.5239	.5279	.5319	.5359
.1	.5398	.5438	.5478	.5517	.5557	.5596	.5636	.5675	.5714	.5753
.2	.5793	.5832	.5871	.5910	.5948	.5987	.6026	.6064	.6103	.6141
.3	.6179	.6217	.6255	.6293	.6331	.6368	.6406	.6443	.6480	.6517
.4	.6554	.6591	.6628	.6664	.6700	.6736	.6772	.6808	.6844	.6879
.5	.6915	.6950	.6985	.7019	.7054	.7088	.7123	.7157	.7190	.7224
.6	.7257	.7291	.7324	.7357	.7389	.7422	.7454	.7486	.7517	.7549
.7	.7580	.7611	.7642	.7673	.7704	.7734	.7764	.7794	.7823	.7852
.8	.7881	.7910	.7939	.7967	.7995	.8023	.8051	.8078	.8106	.8133
.9	.8159	.8186	.8212	.8238	.8264	.8289	.8315	.8340	.8365	.8389
1.0	.8413	.8438	.8461	.8485	.8508	.8531	.8554	.8577	.8599	.8621
1.1	.8643	.8665	.8686	.8708	.8729	.8749	.8770	.8790	.8810	.8830
1.2	.8849	.8869	.8888	.8907	.8925	.8944	.8962	.8980	.8997	.9015
1.3	.9032	.9049	.9066	.9082	.9099	.9115	.9131	.9147	.9162	.9177
1.4	.9192	.9207	.9222	.9236	.9251	.9265	.9279	.9292	.9306	.9319
1.5	.9332	.9345	.9357	.9370	.9382	.9394	.9406	.9418	.9429	.9441
1.6	.9452	.9463	.9474	.9484	.9495	.9505	.9515	.9525	.9535	.9545
1.7	.9554	.9564	.9573	.9582	.9591	.9599	.9608	.9616	.9625	.9633
1.8	.9641	.9649	.9656	.9664	.9671	.9678	.9686	.9693	.9699	.9706
1.9	.9713	.9719	.9726	.9732	.9738	.9744	.9750	.9756	.9761	.9767
2.0	.9772	.9778	.9783	.9788	.9793	.9798	.9803	.9808	.9812	.9817
2.1	.9821	.9826	.9830	.9834	.9838	.9842	.9846	.9850	.9854	.9857
2.2	.9861	.9864	.9868	.9871	.9875	.9878	.9881	.9884	.9887	.9890
2.3	.9893	.9896	.9898	.9901	.9904	.9906	.9909	.9911	.9913	.9916
2.4	.9918	.9920	.9922	.9925	.9927	.9929	.9931	.9932	.9934	.9936
2.5	.9938	.9940	.9941	.9943	.9945	.9946	.9948	.9949	.9951	.9952
2.6	.9953	.9955	.9956	.9957	.9959	.9960	.9961	.9962	.9963	.9964
2.7	.9965	.9966	.9967	.9968	.9969	.9970	.9971	.9972	.9973	.9974
2.8	.9974	.9975	.9976	.9977	.9977	.9978	.9979	.9979	.9980	.9981
2.9	.9981	.9982	.9982	.9983	.9984	.9984	.9985	.9985	.9986	.9986
3.0	.9987	.9987	.9987	.9988	.9988	.9989	.9989	.9989	.9990	.9990
3.1	.9990	.9991	.9991	.9991	.9992	.9992	.9992	.9992	.9993	.9993
3.2	.9993	.9993	.9994	.9994	.9994	.9994	.9994	.9995	.9995	.9995
3.3	.9995	.9995	.9995	.9996	.9996	.9996	.9996	.9996	.9996	.9997
3.4	.9997	.9997	.9997	.9997	.9997	.9997	.9997	.9997	.9997	.9998

Critical values (percentiles) for the F distribution. Upper one-sided 0.05 significance levels; two-sided 0.10 significance levels; 95 percent percentiles.

	Numerator degrees of freedom																		
	1	2	3	4	5	6	7	8	9	10	11	12	13	14	15	16	17	18	19
1	161.4	199.5	215.7	224.6	230.2	234.0	236.8	238.9	240.5	241.9	243.9	245.9	248.0	249.1	250.1	251.1	252.2	253.3	254.3
2	18.51	19.00	19.16	19.25	19.30	19.33	19.35	19.37	19.38	19.40	19.41	19.43	19.45	19.45	19.46	19.47	19.48	19.49	19.50
3	10.13	9.55	9.28	9.12	9.01	8.94	8.89	8.85	8.81	8.79	8.74	8.70	8.66	8.64	8.62	8.59	8.57	8.55	8.53
4	7.71	6.94	6.59	6.39	6.26	6.16	6.09	6.04	6.00	5.96	5.91	5.86	5.80	5.77	5.75	5.72	5.69	5.66	5.63
5	6.61	5.79	5.41	5.19	5.05	4.95	4.88	4.82	4.77	4.74	4.68	4.62	4.56	4.53	4.50	4.46	4.43	4.40	4.36
6	6.99	5.14	4.76	4.53	4.39	4.28	4.21	4.15	4.10	4.06	4.00	3.94	3.87	3.84	3.81	3.77	3.74	3.70	3.67
7	5.59	4.74	4.35	4.12	3.97	3.87	3.79	3.73	3.68	3.64	3.57	3.51	3.44	3.41	3.38	3.34	3.30	3.27	3.23
8	5.32	4.46	4.07	3.84	3.69	3.58	3.50	3.44	3.39	3.35	3.28	3.22	3.15	3.12	3.08	3.04	3.01	2.97	2.93
9	5.12	4.26	3.86	3.63	3.48	3.37	3.29	3.23	3.18	3.14	3.07	3.01	2.94	2.90	2.86	2.83	2.79	2.75	2.71
10	4.96	4.10	3.71	3.48	3.33	3.22	3.14	3.07	3.02	2.98	2.91	2.85	2.77	2.74	2.70	2.66	2.62	2.58	2.54
11	4.84	3.98	3.59	3.36	3.20	3.09	3.01	2.95	2.90	2.85	2.79	2.72	2.65	2.61	2.57	2.53	2.49	2.45	2.40
12	4.75	3.89	3.49	3.26	3.11	3.00	2.91	2.85	2.80	2.75	2.69	2.62	2.54	2.51	2.47	2.43	2.38	2.34	2.30
13	4.67	3.81	3.41	3.18	3.03	2.92	2.83	2.77	2.71	2.67	2.60	2.53	2.46	2.42	2.38	2.34	2.30	2.25	2.21
14	4.60	3.74	3.34	3.11	2.96	2.85	2.76	2.70	2.65	2.60	2.53	2.46	2.39	2.35	2.31	2.27	2.22	2.18	2.13
15	4.54	3.68	3.29	3.06	2.90	2.79	2.71	2.64	2.59	2.54	2.48	2.40	2.33	2.29	2.25	2.20	2.16	2.11	2.07
16	4.49	3.63	3.24	3.01	2.85	2.74	2.66	2.59	2.54	2.49	2.42	2.35	2.28	2.24	2.19	2.15	2.11	2.06	2.01
17	4.45	3.59	3.20	2.96	2.81	2.70	2.61	2.55	2.49	2.45	2.38	2.31	2.23	2.19	2.15	2.10	2.06	2.01	1.96
18	4.41	3.55	3.16	2.93	2.77	2.66	2.58	2.51	2.46	2.41	2.34	2.27	2.19	2.15	2.11	2.06	2.02	1.97	1.92
19	4.38	3.52	3.13	2.90	2.74	2.63	2.54	2.48	2.42	2.38	2.31	2.23	2.16	2.11	2.07	2.03	1.98	1.93	1.88
20	4.35	3.49	3.10	2.87	2.71	2.60	2.51	2.45	2.39	2.35	2.28	2.20	2.12	2.08	2.04	1.99	1.95	1.90	1.84
21	4.32	3.47	3.07	2.84	2.68	2.57	2.49	2.42	2.37	2.32	2.25	2.18	2.10	2.05	2.01	1.96	1.92	1.87	1.81

22	4.30	3.44	3.05	2.82	2.66	2.55	2.46	2.40	2.34	2.30	2.23	2.15	2.07	2.03	1.98	1.94	1.89	1.84	1.78
23	4.28	3.42	3.03	2.80	2.64	2.53	2.44	2.37	2.32	2.27	2.20	2.13	2.05	2.01	1.96	1.91	1.86	1.81	1.76
24	4.26	3.40	3.01	2.78	2.62	2.51	2.42	2.36	2.30	2.25	2.18	2.11	2.03	1.98	1.94	1.89	1.84	1.79	1.73
25	4.24	3.39	2.99	2.76	2.60	2.49	2.40	2.34	2.28	2.24	2.16	2.09	2.01	1.96	1.92	1.87	1.82	1.77	1.71
26	4.23	3.37	2.98	2.74	2.59	2.47	2.39	2.32	2.27	2.22	2.15	2.07	1.99	1.95	1.90	1.85	1.80	1.75	1.69
27	4.21	3.35	2.96	2.73	2.57	2.46	2.37	2.31	2.25	2.20	2.13	2.06	1.97	1.93	1.88	1.84	1.79	1.73	1.67
28	4.20	3.34	2.95	2.71	2.56	2.45	2.36	2.29	2.24	2.19	2.12	2.04	1.96	1.91	1.87	1.82	1.77	1.71	1.65
29	4.18	3.33	2.93	2.70	2.55	2.43	2.35	2.28	2.22	2.18	2.10	2.03	1.94	1.90	1.85	1.81	1.75	1.70	1.64
30	4.17	3.32	2.92	2.69	2.53	2.42	2.33	2.27	2.21	2.16	2.09	2.01	1.93	1.89	1.84	1.79	1.74	1.68	1.62
40	4.08	3.23	2.84	2.61	2.45	2.34	2.25	2.18	2.12	2.08	2.00	1.92	1.84	1.79	1.74	1.69	1.64	1.58	1.51
60	4.00	3.15	2.76	2.53	2.37	2.25	2.17	2.10	2.04	1.99	1.92	1.84	1.75	1.70	1.65	1.59	1.53	1.47	1.39
120	3.92	3.07	2.68	2.45	2.29	2.17	2.09	2.02	1.96	1.91	1.83	1.75	1.66	1.61	1.55	1.50	1.43	1.35	1.25
∞	3.84	3.00	2.60	2.37	2.21	2.10	2.01	1.94	1.88	1.83	1.75	1.67	1.57	1.52	1.46	1.39	1.32	1.22	1.00

Chi-Square Probabilities

df	0.995	0.99	0.975	0.95	0.90	0.10	0.05	0.025	0.01	0.005
1	---	---	0.001	0.004	0.016	2.706	3.841	5.024	6.635	7.879
2	0.010	0.020	0.051	0.103	0.211	4.605	5.991	7.378	9.210	10.597
3	0.072	0.115	0.216	0.352	0.584	6.251	7.815	9.348	11.345	12.838
4	0.207	0.297	0.484	0.711	1.064	7.779	9.488	11.143	13.277	14.860
5	0.412	0.554	0.831	1.145	1.610	9.236	11.070	12.833	15.086	16.750
6	0.676	0.872	1.237	1.635	2.204	10.645	12.592	14.449	16.812	18.548
7	0.989	1.239	1.690	2.167	2.833	12.017	14.067	16.013	18.475	20.278
8	1.344	1.646	2.180	2.733	3.490	13.362	15.507	17.535	20.090	21.955
9	1.735	2.088	2.700	3.325	4.168	14.684	16.919	19.023	21.666	23.589
10	2.156	2.558	3.247	3.940	4.865	15.987	18.307	20.483	23.209	25.188
11	2.603	3.053	3.816	4.575	5.578	17.275	19.675	21.920	24.725	26.757
12	3.074	3.571	4.404	5.226	6.304	18.549	21.026	23.337	26.217	28.300
13	3.565	4.107	5.009	5.892	7.042	19.812	22.362	24.736	27.688	29.819
14	4.075	4.660	5.629	6.571	7.790	21.064	23.685	26.119	29.141	31.319
15	4.601	5.229	6.262	7.261	8.547	22.307	24.996	27.488	30.578	32.801
16	5.142	5.812	6.908	7.962	9.312	23.542	26.296	28.845	32.000	34.267
17	5.697	6.408	7.564	8.672	10.085	24.769	27.587	30.191	33.409	35.718
18	6.265	7.015	8.231	9.390	10.865	25.989	28.869	31.526	34.805	37.156
19	6.844	7.633	8.907	10.117	11.651	27.204	30.144	32.852	36.191	38.582
20	7.434	8.260	9.591	10.851	12.443	28.412	31.410	34.170	37.566	39.997
21	8.034	8.897	10.283	11.591	13.240	29.615	32.671	35.479	38.932	41.401
22	8.643	9.542	10.982	12.338	14.041	30.813	33.924	36.781	40.289	42.796
23	9.260	10.196	11.689	13.091	14.848	32.007	35.172	38.076	41.638	44.181
24	9.886	10.856	12.401	13.848	15.659	33.196	36.415	39.364	42.980	45.559

25	10.520	11.524	13.120	14.611	16.473	34.382	37.652	40.646	44.314	46.928
26	11.160	12.198	13.844	15.379	17.292	35.563	38.885	41.923	45.642	48.290
27	11.808	12.879	14.573	16.151	18.114	36.741	40.113	43.195	46.963	49.645
28	12.461	13.565	15.308	16.928	18.939	37.916	41.337	44.461	48.278	50.993
29	13.121	14.256	16.047	17.708	19.768	39.087	42.557	45.722	49.588	52.336
30	13.787	14.953	16.791	18.493	20.599	40.256	43.773	46.979	50.892	53.672
40	20.707	22.164	24.433	26.509	29.051	51.805	55.758	59.342	63.691	66.766
50	27.991	29.707	32.357	34.764	37.689	63.167	67.505	71.420	76.154	79.490
60	35.534	37.485	40.482	43.188	46.459	74.397	79.082	83.298	88.379	91.952
70	43.275	45.442	48.758	51.739	55.329	85.527	90.531	95.023	100.425	104.215
80	51.172	53.540	57.153	60.391	64.278	96.578	101.879	106.629	112.329	116.321
90	59.196	61.754	65.647	69.126	73.291	107.565	113.145	118.136	124.116	128.299
100	67.328	70.065	74.222	77.929	82.358	118.498	124.342	129.561	135.807	140.169

Other Publications on Mathematics & Statistics

S.No.	Title	Author	ISBN	Year
1	Agricultural Statistics: A Guide for Competitive Examinations	Kushwaha, K.S.	9789381450314	2012
2	Applied Computational Biology and Statistics in Biotechnology and Bioinformatics (Set of 2 Vols.)	Roy, A.K.	9789380235929	2012
3	Applied Statistical Techniques	Imam, Ekwal	9789383305537	2015
4	Applied Statistics for Agricultural Sciences	Venkatesan, D.	9789383305285	2014
5	Basic Concepts in Statistics	Kushwaha, K.S.	9788189422400	2009
6	Computers in Agriculture	Manish K. Sharma & Anil Bhat	9789385516160	2017
7	Emerging Technologies of the 21st Century	Roy, A.K.	9789383305339	2015
8	Inferential Statistics	Kushwaha, K.S.	9789385516443	2017
9	Objective Agriculture Statistics	Kushwaha, K.S.	9789380235615	2011
10	R Statistics	D K Samuel	9789385516146	2017
11	Statistical Designs and Analysis for Agricultural Field Experiments	Katyal, Vijay & D.M.Hegde	9789381450840	2013
12	Statistical Methods for Agricultural Field Experiments	Katyal, Vijay & B.Gangwar	9789380235424	2011
13	The Theory of Sample Surveys and Statistical Decisions	Rajesh Kumar	9788189422899	2009